Patrick Moore's Practical Astronomy Series

For other titles published in the series, go to
http://www.springer.com/series/3192

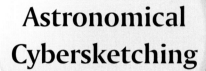

Astronomical Cybersketching

Observational Drawing with PDAs and Tablet PCs

Peter Grego

Springer

Peter Grego

ISSN 1431-9756
ISBN 978-0-387-85350-5   e-ISBN 978-0-387-85351-2
DOI 10.1007/978-0-387-85351-2

Library of Congress Control Number: 2009920356

Printed on acid-free paper

springer.com

To Mike James, teacher of art.

Acknowledgements

My thanks to John Watson, who enthusiastically supported this project from the outset. Among all the hardworking folk at Springer who helped this book along the way, special thanks to Harry Blom and Maury Solomon for their help and patience. I would also like to thank my friends David A Hardy and Dale Holt for their help and input.

Contents

Acknowledgements vii

About the Author. xi

Introduction. xiii

Part I: Hardware: Past, Present, and Future. 1

Chapter One – From Carefully Tooled Gears to Totally Cool Gear 3

Chapter Two – Computers Get Personal. 25

Chapter Three – The Power of the Portable 71

Chapter Four – Handheld Cyberware 85

Chapter Five – Portable Data Storage. 97

Part II: Software and How to Use It 109

Chapter Six – Electronic Skies. 111

Chapter Seven – In Graphic Realms 135

Chapter Eight – Cybersketching Challenges. 167

Glossary 205

Index. 211

About the Author

Peter Grego has been a regular watcher of the night skies since 1976 and began studying the Moon in 1982. He observes from his garden in St. Dennis, Cornwall, UK, using a variety of instruments, ranging from a 100 mm refractor to a 300 mm Newtonian, but his favorite is his 200 mm SCT. Grego's primary interests are observing the Moon and bright planets, but he occasionally likes to 'go deep' during the dark of the Moon.

Grego has directed the Lunar Section of Britain's Society for Popular Astronomy since 1984 and has been the Lunar Topographical Coordinator of the British Astronomical Association since 2006. He edits four astronomy publications: *Luna* (Journal of the SPA Lunar Section), *The New Moon* (topographic journal of the BAA Lunar Section), the *SPA News Circular*, and *Popular Astronomy* magazine. He is also the layout editor for the Society for the History of Astronomy's *Newsletter*.

He has written and illustrated the monthly *MoonWatch* column in *Astronomy Now* magazine since 1997 and is the observing Q&A writer for *Sky at Night* magazine. Grego maintains his own web site at www.lunarobservers.com and is webmaster of the BAA Lunar Section web site at www.baalunarsection.org.uk.

Grego is also the author of 15 books, including *The Moon and How to Observe It* (Springer, 2005), *Venus and Mercury and How to Observe Them* (Springer, 2007), *Moon Observer's Guide* (Philips/Firefly, 2004), *Need to Know? Stargazing* (Collins, 2005), *Need to Know? Universe* (Collins, 2006), and *Solar System Observer's Guide* (Philips/Firefly, 2005). He is a Fellow of the Royal Astronomical Society and a member of the SPA, SHA, and BAA. He has given many talks to astronomical societies around the UK and has been featured on a number of radio and television broadcasts.

Introduction

Sketching the Skies

Suddenly and without warning, a new star appeared in the night sky, and everyone in the community was alarmed. Nobody could remember having seen its like before. Dazzling to look at, this unexpected intruder in the heavenly vault gave off a light that almost rivaled that of the full Moon, drowning out the familiar patterns of stars with its glare. The new star's steady white light penetrated deep into the sacred cave, illuminating an age-old patchwork of intricately drawn pictographs; some of these depicted terrestrial objects and events, from mundane sketches of bison to vast and sweeping panoramic images of wild galloping horses. Other scenes showed celestial phenomena, such as the phases of the Moon and prominent asterisms, or star patterns.

The next morning, accompanied by solemn chanting in which the entire community participated, an elderly shaman entered the sacred cave by the light of a fiery brand and selected a suitable area upon which to depict the new star. Once the artwork was finished, the shaman reappeared at the cave entrance; he held out his arms wide to the slowly brightening morning skies and announced that the powerful magic of the new star had been captured and could now be used to ensure the continuing prosperity of his tribe.

About 30,000 years later, in the same beautiful part of southwestern France, the entrance to the famous world heritage-designated caves at Lascaux was illuminated by another striking celestial spectacle – a piece of midsummer midnight magic which was every bit as compelling to sketch as that shaman of old. Across the fertile plain of the Jurançon, and above the distant silhouetted peaks of the Pyrenees Mountains, a full Moon shared the same section of low southern sky as the planet Jupiter. Unlike our distant ancestor, a torch was not needed to illuminate the artwork. The backlit illuminated screen of a touchscreen handheld computer gave the image a perfect and even illumination; nor was the palette of colors from which to choose limited, or the range of effects to apply to the artwork. The only limiting factors were artistic competence and the amount of artistic license to take with the sketch (Figure 2).

Figure 1. Anyone familiar with the constellations might be tempted to think that this vivid portrayal of the front of a bull, taken from a depiction on the cave walls at Lascaux, represents the constellation of Taurus the Bull, its head and horns marked by the Hyades open cluster. It might even be imagined that the smaller Pleiades star cluster is depicted to its upper right; compare it with a picture of the constellation (Peter Grego).

Of course, the story about the ancient shaman and the unexpected supernova is purely a product of the imagination – at least, its details are – but it is true that our remote ancestors sketched representations of a wide variety of terrestrial and celestial phenomena on the walls of their cave homes and sacred places. The appearance of bright supernovae – stars that explode as they reach the end of their lives – must have been alarming to our superstitious ancestors, to say the least. Some of the celestial depictions were of unexpected spectacles, like the intense blaze of a supernova or the appearance of a brilliant sky-spanning comet; other sights were more predictable, 'routine' heavenly events, such as the rising and setting of the midsummer or midwinter Sun, the patchwork of spots on the Moon's face, and the familiar configurations of certain star patterns. Archaeoastronomers have identified all these celestial representations in cave paintings, petroglyphs, and carvings from various sites around the world that date back many tens of thousands of years to the dawn of humanity.

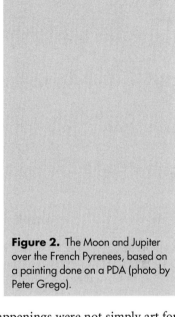

Figure 2. The Moon and Jupiter over the French Pyrenees, based on a painting done on a PDA (photo by Peter Grego).

It is thought that these portrayals of celestial happenings were not simply art for art's sake; they actually served a vital purpose in the minds of the folks who skillfully executed them. Functional representations of the heavens include minutely carved bones and antlers, among which are thought to be records of the Moon's cycles and portrayals of certain constellations. Keeping a tally on the Moon's phases was one way that our ancient ancestors were able to mark the passage of time throughout the year; indeed, many early civilizations based their calendars upon lunar cycles.

Ever since those ancient times, humans have striven to record the world around them and the heavens above in drawings and paintings. Astronomers have been drawing for centuries, ever since the Englishman Thomas Harriot (1560–1621) turned his newly invented little telescope toward the Moon in the summer of 1609 and attempted to sketch the features that were brought into view through his 'optick tube.' Until the invention of photography in the mid-nineteenth century, drawing at the eyepiece was the only way that astronomers could make a visual record of the view through the eyepiece. We have to go back much further in time, long before the telescope was invented, to find the first fairly accurate representations of the starry skies – back to ancient China, where paper maps of the night skies were in use from at least the seventh century CE (Figure 3).

Why Draw?

This author has been a visual observer and an avid at-the-eyepiece sketcher for nearly 30 years. In the following chapters we will argue the case for making observational drawings of a variety of astronomical phenomena. Every amateur

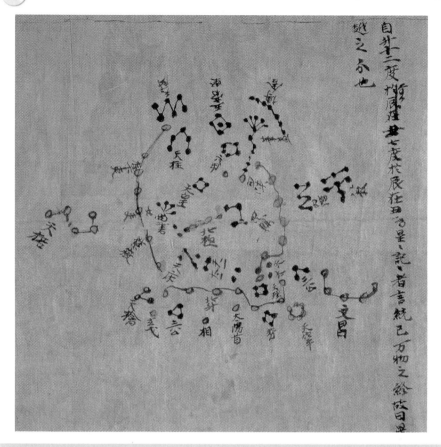

Figure 3. The oldest flat map of the night skies, known as the Dunhuang manuscript, is an ink on paper chart dating back to the seventh century CE. Most of the Chinese constellations depicted are unrecognizable, but the familiar shape of the Plough in Ursa Major is obvious (British Library, London).

astronomer – regardless of his or her own drawing skills – can benefit from being well-versed in various drawing and recording techniques. Drawing is by no means an outmoded and arcane practice. Making observational drawings concentrates and focuses the observer's attention on the subject at hand, enabling the observer to take advantage of moments of good seeing to tease fine, elusive detail out of the image presented in the eyepiece. By attending to the subject in view, those who carefully sketch objects discover more and enjoy observing more than cursory viewers.

Making regular observational drawings improves observing skills all round. When observing and drawing the Moon, with its moving shadows that reveal gloriously detailed topographical scenes of sublime majesty, the lunar landscape is transformed from a confusing jumble of light and dark to a familiar place. After becoming competent at observing and recording planetary detail, what were once tiny bright-colored blobs to the untrained eye become fascinating worlds with ever-changing albedo and cloud variations that can be captured in a drawing. The visual observer who makes the effort to draw deep sky objects will find that nebulae aren't all faint blurry smudges but delicate, subtly detailed wisps of nebulosity whose detail can be sketched.

It's not uncommon to hear disgruntled visual observers complain that drawings can never hope to 'compete' with CCD images. Yet this misses the point. For example, no sane visual observer has ever claimed to have been able to capture all the lunar detail visible through the telescope eyepiece on a single observational drawing. There's simply too much detail discernable on the Moon, even through a small telescope, so visual observers can only hope to produce a drawing that gives a general impression of the appearance of any particular area. No serious visual observer has ever felt that they are in some sort of competition with the image they can see through the eyepiece, so why feel that an image captured by electronic means provides any sort of competition? The root of the problem is a psychological one. Visual observers – especially those with a keen eye and a good drawing ability (either natural or learned) – ruled supreme for more than three centuries. Their drawings were the only means of recording astronomical objects, and amateur astronomers of yesteryear were familiar with books full of drawings by astronomers, rather than high-resolution CCD images. Nowadays, books are full of spectacular full-color CCD images from amateur and professional astronomers; although these are visually pleasing, such images tend to raise the novice's expectations of what they might see through the eyepiece to unrealistic levels, often leading to disappointment and disengagement with observing.

Nevertheless, here are a number of reasons why visual observing is as valid now as it ever was, and will remain valid in the future:

- Drawing is a supremely enjoyable pursuit. If you don't think that you enjoy drawing or were put off drawing by your art teacher at school, give it a go and stick at it for a while.
- Drawings provide a uniquely personal record of observations.
- Attending to detail through drawing allows the observer to concentrate on an object's finer points.
- Drawing enhances every aspect of your observing skills. Making written (or spoken) notes, along with technical aspects of recording features (noting UT, seeing, and other salient details), is also learned in the process.
- Drawing improves your chances of making a scientific discovery.
- Drawing improves your visual skills and enables you to become a better, more accurate observer.

Technological Aids

Now that we have established that drawing is a valid (and perhaps essential) pursuit for the visual observer, let's come to the nitty-gritty of this book – replacing paper and pencil with the computer. We'll call it 'cybersketching' – the prefix *cyber* referring to all things electronic.

Computers and digital imaging technology can do nothing but help the visual observer in many important respects. We shouldn't regret that the so-called Golden Age of visual observation is long gone; that era began to fade away when astrophotography came along and had all but disappeared during the Space Age. We have now entered a more exciting age of visual observation, where we can learn from the great telescopic observers and apply new technology to our hobby.

These days, people are increasingly 'digitizing' their lifestyles, and an increasing number of amateur astronomers of the future will not willingly put up with damp paper and smudgy pencils while juggling with a red torch at the eyepiece. This is where modern technology – in the form of PDAs (personal digital assistants, or handheld computers), UMPCs (ultra-mobile portable computers), and tablet PCs (flat, lightweight, touchscreen portable computers) – provides some neat solutions.

So, what makes PDAs, UMPCs, and tablet PCs so special? Well, for a start, they carry their own source of illumination. This is a big bonus because they obviate the need for a separate source. Computers are supremely versatile, as images can be stored and retrieved, zoomed-in on, modified, and enhanced at will. Most good drawing programs for PDAs and tablet PCs allow the properties of the stylus stroke to be modified in terms of its thickness, shape, intensity, color, texture, and transparency, so that a range of pencil, pen, brush strokes, and other artistic media (such as spray cans, paint rollers) can easily be replicated. The experience is fairly intuitive, in that the user is inputting data onto a screen using a stylus; because it's very much like drawing onto paper with a pencil, using a stylus requires little special skill or expertise to get the hang of. Indeed, virtual sketching is in many ways easier than 'real' sketching, and it can provide a more pleasant experience. First-time users find it a remarkable experience – a kind of 'eureka' moment that one can compare to seeing Saturn through a telescope eyepiece the first time.

This book outlines the techniques involved in astronomical cybersketching – making observational sketches and more detailed 'scientific' drawings of a wide variety of astronomical subjects using modern digital equipment, specifically PDAs, UMPCs, and tablet PCs. Various items of hardware and software are discussed, although with such an ever-growing range of products available on the market, the discussion is necessarily kept to its essentials. Once observational drawings are made at the eyepiece, we move on to deal with the process of producing finished or enhanced drawings at the user's main PC.

Contrary to 'assimilating the masses' in a mundane digital world, new technology can really only serve to liberate people by expanding their knowledge and unleashing the potential of their creativity. As astronomical cybersketching gains in popularity over the coming years, it will produce graphic works whose makers' individuality is as apparent and palpable as that in physical artworks. Hopefully, this modest book will help to further that end.

It is hoped that the techniques revealed in this book encourage many people to try cybersketching for themselves. Whether it represents the future of making records of visual observations of astronomical subjects remains to be seen. Having taken computer in hand into the field, it is hard to imagine what might induce someone to return to the exclusive use of pencil and paper.

Part I

Hardware: Past, Present, and Future

CHAPTER ONE

From Carefully Tooled Gears to Totally Cool Gear

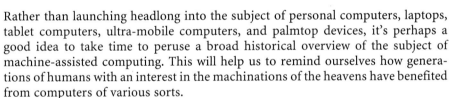

Rather than launching headlong into the subject of personal computers, laptops, tablet computers, ultra-mobile computers, and palmtop devices, it's perhaps a good idea to take time to peruse a broad historical overview of the subject of machine-assisted computing. This will help us to remind ourselves how generations of humans with an interest in the machinations of the heavens have benefited from computers of various sorts.

Ancient cultures in all corners of the globe developed an amazing variety of cosmologies that granted the untouchable occupants of the heavens – the Sun and Moon, the stars, and the planets – a variety of supernatural powers over Earth and over the affairs of humanity. It appeared perfectly clear that the sky gods had their own unique personalities; no two looked the same, and each moved through the sky at its own speed and in its own special way. Most powerful among the sky gods were the Sun and the Moon, which were sometimes interpreted as an endlessly competing pair of deities because of the phenomena of solar and lunar eclipses.

Our ancestors might have imagined that it was possible for mere mortals to understand the intentions of the sky gods if their movements and phenomena were carefully noted. It followed that careful observation over extended periods of time enabled distinct patterns to be recognized, allowing some events to be predicted in advance. Having the ability to predict celestial phenomena gave the watchers of the skies great power, as events on Earth were thought to be affected by them and thus possible to predict as well. This was a pretty handy skill to possess for any ruling class.

P. Grego, *Astronomical Cybersketching*, Patrick Moore's Practical Astronomy Series, DOI 10.1007/978-0-387-85351-2_1, © Springer Science+Business Media, LLC 2009

Archaeological evidence tells us that humans have been systematically watching the skies and recording celestial events for at least 6,000 years. The remains of megalithic constructions such as Stonehenge, whose stones are aligned with specific celestial points, such as the rising and setting points of the midsummer/ midwinter Sun, are impressive evidence of how important people once thought it was to maintain an awareness of celestial phenomena (Figure 1.1).

Great leaps forward in understanding the workings of the cosmos took place in ancient Greece, where philosophers used their intellects to define the universe in

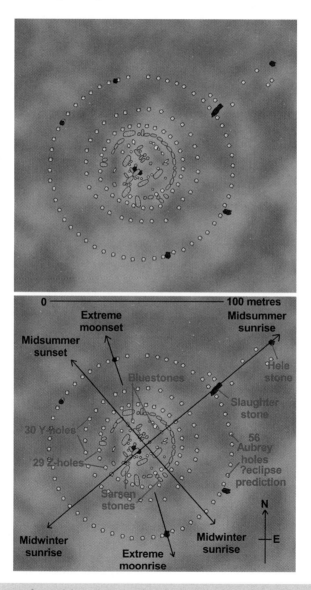

Figure 1.1. Rising from Salisbury Plain in southern England, the ancient, intricately arranged carved rocks of Stonehenge are thought to be some form of astronomical computer (photo by Peter Grego).

physical terms. In the fourth century BCE the philosopher Eudoxus of Cnidus (410–355 BCE) devised a complete system of the universe; in his model Earth lay at the very center of a series of concentric crystal spheres upon which were fastened the Sun, Moon, individual planets, and stars. As for the scale of the universe, mathematics and geometry proved invaluable tools with which to understand realms that were unimaginably vaster than our own planet. Careful observation combined with trigonometry enabled Eratosthenes of Cyrene (276–194 BCE) to accurately measure the circumference of Earth around 240 BCE.

A century later, Hipparchus (190–120 BCE) made the first fairly accurate determinations of the distance and size of the Moon. At around the same time Ptolemy took care to compile an encyclopedia of ancient Babylonian and Greek knowledge, including a definitive atlas of the stars – no fewer than 1,022 of them, contained within 48 constellations. Expanding on Eudoxus's idea of an Earth-centered universe, Claudius Ptolemaeus (around 83–161 CE) explained that the odd looping motions (called 'retrograde motion') of the planets at some points along their paths were produced when the planets performed smaller circular movements, or 'epicycles,' as they orbited Earth. Although the idea of epicycles answered a lot of problems and appeared to explain the clockwork of the cosmos, careful observation over extended periods of time was later to prove their downfall – but the story of modern astronomy and the heliocentric (Sun-centered) universe is far beyond the scope of this chapter.

The Antikythera Mechanism

Around the year 150 BCE, a cargo ship plying the waters near the little island of Antikythera, half way between mainland Greece and Crete, met with disaster. For some reason unknown to us – probably the result of a sudden storm – the vessel capsized and sank some 50 m to the bottom of the Kithirai Channel, where it remained along with its cargo to gather the usual organic submarine exoskeleton until it was discovered almost 2,000 years later. Shipwrecks are not uncommon in this part of the world, as trading between the myriad of islands in the region has been going on since time immemorial. It is thought that this particular ship was laden with loot, en route from the island of Rhodes to the burgeoning city of Rome. Small items soon began to be recovered from the wreck by sponge divers; among the concreted debris, which included fragments of pottery, sculptures, and coins, several items appeared that were markedly different from anything that had been previously found at any archaeological site of such antiquity.

Close examination revealed the fragments of a heavily encrusted, corroded, geared device measuring around 33 cm (13 in.) high, 17 cm (6.7 in.) wide, and 9 cm (3.5 in.) thick (Figure 1.2). Constructed of bronze and originally contained within a wooden frame, the device was engraved with a copious text (more than 3,000 characters in length), which appears to be the device's operating manual. With references to the Sun and Moon, along with the motions of the planets Aphrodite (Venus) and Hermes (Mercury), it is thought that the instrument could have been used to predict various astronomical cycles, such as the synodic month (the interval between full moons) and the metonic cycle (235 lunar months between exact phase repetitions) along with some of the phenomena displayed by the inferior planets. As such, this amazing piece of engineering represents the first portable, programmable computer, demonstrating that the ancient Greeks were far more technologically advanced than they are sometimes given credit for.

Figure 1.2. The main fragment of the Antikythera mechanism, on display at the National Archaeological Museum of Athens. Despite its condition, the great complexity of the device can clearly be seen. Credit: Marsyas, Wikimedia Commons.

Deus Ex Machina

Mechanical devices have always formed part of the astronomer's armory; since mathematics is essential to understanding and predicting celestial events, the abacus was among the first such mechanical devices because it made arithmetic a great deal easier. Abaci were used in ancient Sumeria more than 4,000 years ago, and the earliest Greek abacus in existence has been dated to 300 BCE.

In pre-Columbian Central America, from around 1,000 BCE, the complicated calculations involving the 260-day festival calendar was made easier by the use of calendar wheels. The festival calendar, known as a tzolkin, was based on physical objects, animals, and deities, and it revolved around the numbers 20 (the digits of the 'whole person') and 13 (in their philosophy there were 13 directions in space). Rotations of meshed wheels of 20 and 13 spaces enabled each day to be associated with a different object, and the whole cycle with respect to the 365-day solar calendar repeated itself every 52 years. Calendar wheels were therefore useful for planning events and for telling the future.

Simple naked-eye cross-staffs enabling the measurement of celestial angles have been used since antiquity. More complicated astronomical instruments that permitted calculations to be made in advance included the planisphere and the astrolabe, both of which first appeared in ancient Greece. Consisting of a map of the stars and an overlay that could be rotated to approximate the position of the horizon at any given date and time, the planisphere is an elegant, though rudimentary, device that allows the operator to calculate the rising and setting times of the Sun and stars and their elevation above (or below) the horizon at any given time. Planispheres are still beloved by amateur astronomers; indeed, most modern astronomical computer programs contain a facility to create a planisphere display. Astrolabes are a potent combination of the planisphere and a sighting device called a dioptra; thought to have been invented by Hipparchus, astrolabes permitted calculations to be made on the basis of observations, enabling numerous problems in spherical astronomy to be solved. Perhaps the most prolific and proficient exponents of the astrolabe were astronomers of the medieval Islamic world, where they were employed for astronomy, navigation, and surveying, in addition to being put to use as timekeepers for religious purposes.

Planispheres and astrolabes were used extensively by astrologers in medieval Europe to construct horoscopes (Figure 1.3). Although we now know that astrology is pseudoscience, without any scientific merit, there was no shortage of eminent

Figure 1.3. A superb brass astrolabe manufactured by Georg Hartmann in Nuremberg in 1537, now in the Scientific Instruments Collection of Yale University (Ragesoss, Wikimedia Commons).

practitioners in the West who combined astrology with their more serious astronomical pursuits. For example, Johannes Kepler (1571–1630), brilliant mathematician and originator of the laws of planetary motion, was convinced of the merits of astrology and devised his own system based upon harmonic theory. Some 800 horoscopes formulated by Kepler are still in existence, and certain lucky predictions for the year 1595 – including foretelling a peasants' revolt, forebodings of incursions by the Ottoman Empire in the east, and predictions of a spell of bitter cold – brought his astrological talents into great renown.

Figure 1.4. A celestial globe and a copy of Adriaan Metius' book *Institutiones Astronomiae Geographicae* feature in Johannes Vermeer's painting *The Astronomer* (1668). The book is open at Chapter Three, where it is stated that along with knowledge of geometry and the aid of mechanical instruments, there is a recommendation for 'inspiration from God' for astronomical research. Nowadays many amateurs echo this sentiment by praying that the battery on their laptop or PDA holds out during a night's observing session.

Lookers and Optick Tubes

In most textbooks on astronomy credit for the invention of the telescope goes to the Dutch–German lens maker Hans Lippershey (1570–1619) of Middelburg, Zeeland, in the Netherlands. One version of the traditional story says that children playing in his workshop stumbled upon the fact that the combination of a negative (concave) and a positive (convex) lens will magnify a distant image, provided that the negative lens is held near the eye and the lenses are firmly held at the right distance from each other; why Lippershey's children would be allowed to play in his workshop full of delicate and expensive glass items is not explained, and of course the story is utterly unverifiable. Regardless of whether the discovery was made by accident or by careful experiment, Lippershey presented his invention – a device that he called a kijker (a 'looker,' which magnified just three times) – to the Dutch government in October 1608, with the intention of obtaining a patent, stating that such an instrument would have enormous military potential. However, it was thought that there was little chance of successfully keeping the invention a secret or preventing others from making their own telescopes, and the patent was declined. Nevertheless, Lippershey was well rewarded for his design, and he went on to make several binocular telescopes for the government.

Two other Dutch opticians later claimed to have come up with the idea of the telescope prior to Lippershey – Jacob Metius (1571–1628) of Alkmaar in the Northern Netherlands, who actually filed his patent application just a few weeks after Lippershey, and the notorious counterfeiter Sacharias Jansen of Lippershey's hometown of Middelburg, who claimed to have made a telescope as early as 1604. However, Lippershey's patent application represents the earliest known documentation concerning an actual telescope, so the credit rightly remains with Lippershey. Interestingly, the surnames of all three pioneering opticians have been given to craters on the Moon – Lippershey, a fairly insignificant 6.8 km diameter pit in southern Mare Nubium (Sea of Clouds); Jansen, an eroded 23 km crater in northern Mare Tranquillitatis (Sea of Tranquillity); and 88 km diameter Metius in the southeastern corner of the Moon (although the latter is actually named after Jacob's brother, Adriaan).

Before proceeding to the telescopic era, it's worth pointing out that the first and only known pre-telescopic map of the Moon based on naked-eye observations was made by the Englishman William Gilbert (1544–1603) in the early seventeenth century (Figure 1.5). Gilbert's drawing is by no means the most detailed depiction of the lunar surface, nor is it the most expertly drafted. In fact, one of the most prominent ink strokes on it is an obvious error in positioning. The map is, however, unique in that it designated a set of names to the Moon's features. Among the quaint nomenclature on Gilbert's map is 'Britannia,' designating the dark oval patch we now call Mare Crisium (Sea of Crises) and 'Insula Longa' (Long Island) for the region now called Oceanus Procellarum (Ocean of Storms). It's true that the great Leonardo da Vinci (1452–1519) sketched the Moon way back in the early sixteenth century – and the fragment showing the eastern half of the Moon extant today is a good representation – but he was content not to name the features he saw (Figure 1.6). Leonardo considered the Moon an Earth-like world, the bright areas representing seas and the dark areas continents (contrary to common belief at the time, which imagined that the dark areas were seas). Leonardo also correctly deduced the true cause of Earthshine – that faint illumination of the dark side of the Moon when it is a crescent phase – ascribing it to sunlight reflected onto the Moon from Earth. Still, it's incredible that Gilbert's map is the only surviving pre-telescopic lunar map – incredible, considering that the Moon displays such obvious features when viewed with the average unaided eye.

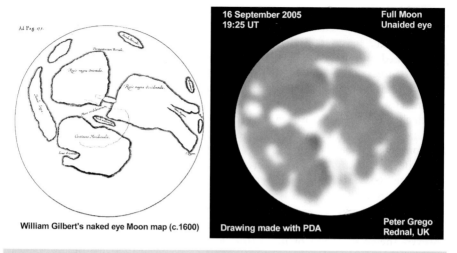

Figure 1.5. William Gilbert's naked-eye map of the Moon, circa 1600, compared with a sketch by the author using a PDA in 2005.

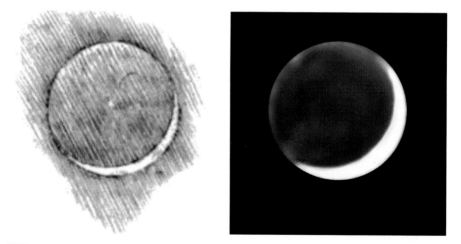

Figure 1.6. a Leonardo da Vinci's drawing of the Moon, made around 1500. Leonardo may have depicted the other half of the Moon, but the drawing has never been located. **b** Leonardo da Vinci's sketch of an Earthshine-illuminated young crescent Moon compared with a sketch of the Moon at a similar phase by the author using a PDA.

In the year 1609, news of the invention of the telescope and the principles involved in its construction rapidly spread throughout Europe. A series of astonishing observational discoveries by Galileo Galilei (1564–1642), commencing in late 1609, ushered in the era of observational astronomy. Squinting through a homemade instrument barely more sophisticated than a child's toy telescope of today, Galileo observed that the surface of the Moon displayed majestic highland regions packed full of craters and mountains, along with vast smooth gray plains pockmarked by numerous bright spots surrounded by light streaks. In his book *Sidereus Nuncius* (*The Starry Messenger*, 1610) the scientist wrote, '. . . the surface of the Moon is neither smooth nor uniform, nor very accurately spherical, as is assumed by a great many philosophers . . . it is uneven, rough, replete with cavities and packed with protruding eminences.'

Galileo's original inkwash drawings of the Moon show that he was an accurate and competent artist. His telescope was not powerful enough to discern any detail on Jupiter's disk, although he did find that the giant planet was attended by four large satellites; in fact, we still call Io, Europa, Ganymede, and Callisto the 'Galilean moons.' Among numerous other major astronomical discoveries, he saw that the surface of the Sun occasionally displayed sunspots, that Saturn was surrounded by mysterious 'appendages' (later found to be a ring system), and resolved the misty band of the Milky Way into countless faint stars. In 1612 Galileo also noted the planet Neptune in the same field of view as Jupiter but thought that it was merely a background star. Hence, this most distant of the major planets had to wait until 1846 to be discovered by Urbain Le Verrier, Johann Galle, and Heinrich d'Arrest.

The Clockwork Universe

In the Age of Enlightenment, humanity was gradually beginning to feel at ease with the fact that Earth was a relatively small globe in orbit around the Sun – one planet of many – rather than being an immovable rock anchored at the very hub of the universe. From the time of Galileo up until the telescopic discovery of Uranus by William Herschel in 1781, the known Solar System extended out as far as the orbit of Saturn.

Not content with having those fortunate enough to own a telescope able to appreciate the wonders of the Solar System, the clockmaker George Graham (c. 1674–1751) decided to build his own small mechanical Solar System for the purpose of instructing and enlightening its viewers. With Graham's consent, his first model Solar System was copied by the instrument maker John Rowley, who made one for Prince Eugene of Savoy and another for his patron Charles Boyle, 4th Earl of Orrery. This mechanical marvel – now known as an 'orrery' – demonstrated the orbits of the planets around the Sun, their axial rotation, and the orbits of their satellites, all in the correct ratio of speed (Figure 1.7). The finest orreries could be used as computers to determine the positions of the planets with respect to each other at any time in the past or future, although their predictive accuracy fell with the interval of time before or after the date to which it was originally set. For example, an orrery might be geared so that 12 Earth rotations around the Sun matched one of Jupiter's; however, Jupiter's orbital period is in fact 11.86 years long, so after several orbits there would be a big discrepancy between the actual and predicted position of Jupiter with respect to Earth.

Figure 1.7. A small eighteenth-century orrery showing the inner Solar System (Kaptain Kobold, Wikimedia Commons).

Making a Difference

A calculating machine that performed basic arithmetical functions was built as long ago as 1642 by the French mathematician Blaise Pascal (1623–1662); the device was made to assist the work of Pascal's father, who had the very important job of king's commissioner of taxes in Rouen. In the late seventeenth century Gottfried Wilhelm von Leibniz invented a computer capable of similar functions, but it was not until the early nineteenth century that successful mechanical calculators capable of addition, subtraction, multiplication, and division made their appearance.

It's astonishing to think that much of the routine mathematical work that put people into orbit around Earth and landed astronauts on the Moon in the 1960s was performed using an unassuming little mechanical analog computer – the 'humble' slide rule. Consisting of several logarithmically divided rules and a sliding cursor, slide rules can be used for multiplication and division, in addition to performing more complex roots, logarithms, and trigonometric functions. A basic slide rule was invented by William Oughtred (1575–1660) in 1622, but the modern version dates back to an 1850 design by Amedee Mannheim (1831–1906).

Frustrated with inaccuracies introduced in manually calculated tables, the English mathematician Charles Babbage (1791–1871) realized that it was possible to minimize errors by automating the process of making calculations. In an 1822 paper to the Royal Astronomical Society entitled *Note on the application of machinery to the computation of very big mathematical tables* Babbage proposed a hand-cranked calculating machine which used the decimal system. He was confident that advanced models of such calculating machines, which he described as 'difference engines,' could be powered by steam and be capable of automatically calculating and printing astronomical tables with minimal human input. After 10 years of working on the government-backed project (which sadly never came to fruition), Babbage abandoned it to work on his ideas for a far grander project – a programmable, automated, mechanized digital computer which he termed an 'analytical engine.' Ultimately proving too advanced for their day, many of Babbage's ideas and concepts were not realized in practice until more than a century later with the arrival of electronic computers. In 1991 the London Science Museum successfully constructed a difference engine based on Babbage's original designs; it was clearly demonstrated that Babbage's device was indeed fully functioning and capable of performing complex calculations.

Computers with Punch

Despite the promise of faster and more accurate results being offered by mechanical computers, humans were to remain responsible for most of the computations of astronomical tables and mathematical problem solving throughout the nineteenth century. However, observations were becoming increasingly refined, and there was a need for greater accuracy; moreover, there was an ever-growing volume of observational data to analyze. Because of this, many astronomical research institutions found themselves compelled to employ more people to perform the necessary calculations to assist the professional astronomers.

Some of these human computers achieved notable successes. At Harvard College Observatory a team of them was employed by Edward Charles Pickering (1846–1914) to measure and catalog stellar brightnesses in the observatory's photographic collection. Based on this work, one computer, Henrietta Swan Leavitt (1868–1921), discovered the luminosity–period relationship of Cepheid variable stars – an immensely important finding that allowed interstellar and near-intergalactic distances to be determined with great accuracy. Other notable human computers at Harvard included Annie Jump Cannon (1863–1941) and Antonia Maury (1866–1952), both of whom worked on the classification of stellar spectra.

At the US Naval Observatory, human computers had been a growing component of the staff since mid-century, employed to calculate tables for the *American Ephemeris and Nautical Almanac*. Notable among these number crunchers was Maria Mitchell (1818–1889), who calculated tables of the positions of Venus; in 1847 Mitchell achieved worldwide fame as the discoverer of comet C/1847 T1, the first woman to discover a comet since Caroline Herschel (1750–1848) exactly 50 years earlier. In the ensuing era of aircraft travel in the twentieth century, the *American Ephemeris and Nautical Almanac* contained aeronautical supplements. Eventually the stand-alone *Air Almanac* was published, which contained a number of astronomical tables and was the first ephemeris to be produced with the aid of punched card computers.

Pioneered in the United States, first by the US Census Bureau and later by companies such as International Business Machines (IBM) and Remington, punched card computers read (and produced results on) resilient cardboard sheets containing patterns of holes arranged in rows and columns; each hole's location represented a certain value, and the cards could be automatically read. Punched cards had been used for decades in the textile industry to program the patterns produced by looms – in fact, this was the inspiration behind Hermann Hollerith's original US census machine of 1890 – and they were (and still are) used in those quaint old fairground organs to produce a constant stream of music with no human input.

In 1937 Columbia University and IBM established the Thomas J. Watson Astronomical Computing Bureau, in collaboration with Wallace Eckert (1902–1971). Eckert detailed the venture in his paper *Punched card methods in scientific computation* (1940). At the beginning of the Second World War, Eckert took on the directorship of the US Almanac Office and became head astronomer at the US Naval Observatory. Eckert quickly automated many of the computational processes that had been laboriously performed manually for many decades. He introduced punched card technology to compute and print astronomical tables, and the first publications to make use of the new technology were the *Nautical Almanac* and the *American Air Almanac* of 1940. Eckert was later instrumental in constructing several astronomical computers, notably the Selective Sequence Electronic Calculator (SSEC) in 1949 and the Naval Ordnance Research Calculator (NORC) in 1954 (see below).

Punched card computers were in use for collecting and arranging large volumes of data until relatively recently – for example, the ticket to the lending library of Birmingham Central Library in the late 1970s was a small plastic punched hole affair. Doubtless these are used in some odd places to this day. Electrically powered punched card computers were capable of sifting through large volumes of data. In comparison to the computers of today, these devices operated at glacially sluggish speeds; each punched card contained around 80 characters, and a processing rate of one card per second was typical. Electromechanical punched card machines grew in popularity during the early twentieth century, and they were to perform much of the world's business and scientific computing until the invention of electronic computers.

Electronic Brains

Among the earliest electronic computers were those built to help the Allied war effort during the 1940s. During the Second World War, at the top secret code breaking research facility at Bletchley Park in southern England, a series of electronic computers called Colossus were designed by engineer Thomas Flowers (1905–1998) with input from mathematician Max Newman (1897–1984) and Alan Turing (1912–1954), among a host of other brilliant mathematical minds. Ten Colossi had been constructed by the end of the war. The first Colossi were used to crack the Nazi Lorenz cipher machine system, using more than 1,500 electronic valves to simulate the workings of the supposedly unassailable code machine's mechanical rotors. After the war two Colossi were taken to GCHQ to be used for code breaking purposes during the Cold War. Eventually all the machines were dismantled and destroyed, along with the plans and drawings for the original Colossi – the sad consequence of the United Kingdom's institutionally overly secretive culture. Thankfully, a group of dedicated people led by Tony Sale built a working replica of Colossus after extensive consultation with many of the people involved in Colossus (including Flowers himself); this incredible machine is open to the public at Bletchley Park.

In the United States, a larger general purpose electronic digital computer was designed by John Eckert (1919–1995, no relation to Wallace Eckert, above) and John Mauchly (1907–1980). Known as ENIAC (Electrical Numerical Integrator And Calculator), it was first used to calculate firing tables for the US Army Ballistic Research Laboratory and was operational until 1955 (Figure 1.8). ENIAC was by no means a small device – weighing 30 metric tons it contained 18,000 vacuum tubes, occupied around about 170 m^2, and required about 180,000 W of electrical power to operate. Someone charmingly described the machine as having a computational power 'faster than thought.'

Figure 1.8. ENIAC in operation at the Moore School of Electrical Engineering at the University of Pennsylvania (US Army).

Both Colossus and ENIAC were hard-wired machines used to solve specific problems. Neither could switch from one program to another without being rewired according to the mathematical problem being addressed. This limitation prompted John von Neumann (1903–1957) to suggest that computers of the future would be far faster, more flexible, and efficient if they had a fairly simple physical structure and were capable of performing any computational task using a programmed control; importantly, the physical wiring of the computer need not be altered to achieve this. Stored-program computers with a von Neumann-type architecture (modern computers are among them) operate according to internally stored instructions; they are capable of creating and storing further instructions as required, and they subsequently execute those instructions. The computer's memory serves to rack up and assemble all the various parts of a long computation. In addition, instructions can be amended during a computation, making these computers extremely versatile.

Such machines – recognized as the first generation of modern computers – were created in the post-war era. Random access memory (RAM) was introduced, enabling any particular item of stored information to be accessed at any time. One of the first was the Small-Scale Experimental Machine (SSEM, nicknamed 'Baby'), built at the University of Manchester in England, which ran its first program in June 1948; a working replica of Baby is on show at the Manchester Museum of Science and Industry. In the following year, the full-sized Manchester Mark 1 computer (also known as the Manchester Automatic Digital Machine, or MADM) came into operation, which was used for scientific computation at the university. It was also hired out for computational work for the jaw-dropping regal sum of £5,000 per hour (a veritable fortune in those days, and no mean sum today). With its 4,000 valves, MADM incorporated a high-speed magnetic drum (an early form of hard drive) as an adjunct to data storage, and the machine served as the basis for the world's first commercially available computer, the Ferranti Mark 1.

IBM's Selective Sequence Electronic Calculator (SSEC), constructed under the direction of Wallace Eckert and his Watson Scientific Computing Laboratory staff in the late 1940s, found a home at IBM's headquarters building in Manhattan, where it took up 200 m^2 of wall space in a large U-shaped room. Visible to passers-by on Madison Avenue, the computer became a well-known landmark and even featured in a humorous cartoon on the cover of *The New Yorker*. A limestone plaque dedicated the computer: 'This machine will assist the scientist in institutions of learning, in government, and in industry to explore the consequences of man's thought to the outermost reaches of time, space, and physical conditions.'

Eckert later directed the design and construction of the impressive Naval Ordnance Research Calculator (NORC), whose wiring was flushed with its first sparks of calculative activity in 1954. Eckert used NORC to compute lunar ephemerides using the equations of Ernest Brown (1866–1938). So accurate were the results that the presence of mascons (concentrations of mass) near the Moon's surface was inferred, and Eckert's *Improved Lunar Ephemeris* was used by the *Apollo* lunar landing missions. Eckert wrote, 'A calculation involving a billion arithmetical operations on large numbers can be completed on the NORC in approximately one day, yet more powerful calculators are foreseen to meet the ever-increasing demands of science and technology where the solution of a large problem generates even larger problems.' It's a sobering thought that modern supercomputers are more than a billion times faster than NORC, which was the world's most powerful computer until 1963. NORC remained operational until 1968.

Several important technological advances accelerated the speed of computer development, notably the invention of the transistor in 1947 and the invention of magnetic core memory in 1951. Transistors are small semiconductors that can switch and modulate electric current. They became an effective replacement for vacuum tubes, which, being somewhat akin to light bulbs, were bulky, costly, immensely less reliable, and required the machines containing them to receive a great deal of maintenance. From the 1950s, computers used for big number crunching purposes began to appear all around the globe, in universities, research institutions, governmental departments, and businesses.

The Birth of Computer Graphics

In the mid-twentieth century the world of art also experienced its first encounters with electronically generated graphics. In 1950 the American mathematician and artist Ben Laposky (1914–2000) created the first computer graphics by generating images using an electronic analog computer, capturing the flowing wave forms that were produced on an oscilloscope's screen using high-speed film. Laposky called these electronic abstractions 'oscillons,' and they look very much like the images produced at random by the Windows *Mystify* screen saver.

Increasing commercial and military air traffic following the Second World War demanded the development of a far more serious early use of computers and computer-generated graphics – those deployed by radar systems. Tasked by the US Air Force to develop a way to protect against air attacks from hostile powers, MIT's Lincoln Laboratory developed the SAGE (Semiautomatic Ground Environment) air defense system. SAGE used dozens of Whirlwind computers in a network of radar stations to plot blips on CRTs; the blips represented incoming aircraft based on radar gathered data. Operators were able to apply a gun-shaped light pen to each blip to bring up additional text information about the target on-screen, such as its identification, airspeed, and direction. First developed in 1955, the system represented one of the first practical uses of computer graphics (Figure 1.9).

In 1957 the US National Bureau of Standards produced the first computer-processed photograph (a drum-scanned image), and a year later John Whitney used an analog computer – an M-5 anti-aircraft gunsight computer – to produce simple animations, including the psychedelic *Catalog*. In 1961 Ivan Sutherland of the Massachusetts Institute of Technology (MIT) began to develop his revolutionary *Sketchpad* vector art software, working on it between 3 and 6 am for more than a year – odd hours, but the machine was used for air defense purposes during the rest of the day. Released in 1963, *Sketchpad* was announced as 'a man–machine graphical communication system' and was run on an MIT Lincoln Laboratory TX-2 computer using a light pen to manipulate the graphics.

Simple animations aside, the first true computer-generated film is considered to have been made in 1963 by Edward Zajac of Bell Laboratories using an IBM 7090 mainframe computer. Titled *A Two Gyro Gravity Gradient Altitude Control System*, the film demonstrated that an orbiting satellite could be stabilized to keep one face permanently turned Earthward. In the same year the first computer

Figure 1.9. SAGE in operation. A USAF radar operator applies a light pen to the screen to bring up information about the displayed target (USAF).

game appeared, called *Spacewar!* Created by Steve Russell and colleagues working on a DEC Digital PDP-1 at MIT, its concept is familiar to video games players today – a shoot 'em up involving two spacecraft in the vicinity of a star whose gravity affects the ships (Figure 1.10).

A collaboration between General Motors and IBM saw the development of the DAC-1 (Design Augmented by Computers) system in 1964, heralding the first industrial use of computer-aided design (CAD). Using DAC-1, a 3D model of an automobile could be constructed in the computer and displayed on-screen in wireframe form at any desired angle and rotation. Boeing quickly followed suit; indeed, the very term 'computer graphics' was first used by Boeing employee Verne Hudson. One of the company's first uses of CAD was to design the 'Boeing Man,' a wireframe representation of a human being, used for ergonomic purposes. It goes without saying that CAD is now considered an indispensable tool in every major engineering endeavor.

In the ensuing decades, large companies and organizations gradually twigged onto the idea that computer graphics and animations were going to be among the 'next big things' to impress an increasingly sophisticated general public and to sell their products. In 1982 the Walt Disney Studios released *Tron*, a movie that was groundbreaking for its extensive use of computer graphics. Fascinated by video games, the movie's writer and director Steven Lisberger was inspired to

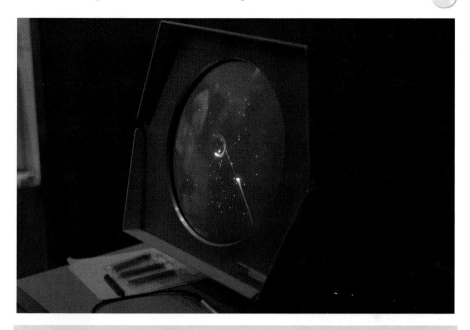

Figure 1.10. *Spacewar!* The world's first computer game, demonstrated on the monitor of the only remaining operational DEC PDP-1 computer, at the Computer History Museum in Mountain View, California (Joi Ito, Wikimedia Commons).

produce a movie that brought the world of computers to a general audience. He succeeded in producing a distinctive-looking film that combines live action with computer simulation; the film remains eminently watchable to this day. The computer animation sequences of *Tron* were produced by four of the leading computer graphics companies of their day, including Information International Inc., of Culver City, California, who owned the 'Super Foonly F-1,' the fastest PDP-10 mainframe computer ever made. The success of *Tron* demonstrated that computers and computer graphics were capable of entertaining everyone – not just nerds and video arcade loungers – and it wasn't long before the viewers of both large and small screens were treated to the best that the embryonic computer graphics industry could muster.

One notable proponent of computer graphics and animated simulations was the Computer Graphics Laboratory (CGL) at the Jet Propulsion Laboratory (JPL) in Pasadena, California, which was established by Bob Holzman in 1977. The CGL was charged with the task of visualizing, in both still and animated forms, space-related themes and various data being returned by NASA space missions. Among the first space missions to benefit from the CGL's computer wizardry were the two *Voyager* space probes, which had been launched in 1977 to conduct a grand tour of the outer Solar System. Some fascinating graphics of the *Voyager* probes' encounters with Jupiter and its moons (1979) and the magnificent Saturn system (1980 and 1981) were created (Figure 1.11). *Voyager 2* went on to fly by Uranus (1986) and Neptune (1989).

Figure 1.11. A computer-generated wireframe graphic animation, and the artwork based upon it, by Jim Blinn of the JPL CGL. The image shows the *Voyager 2* space probe's position on August 25, 1981 (NASA).

Memories of the Space Age

Magnetic core memory, an early form of RAM, was invented by Jay W. Forrester of MIT in 1951. It consisted of tiny magnetic ceramic rings suspended within a network of threaded wires; the computer could change or retrieve the information from each ring (this information being represented by a 1 or a 0, according to its polarity). Core stacks consisting of hundreds of thousands of bytes of memory could be built up in this way, producing computers with the ability to get to grips with some serious space age problems. Magnetic core memory held sway during the 1950s and 1960s, and the technology was instrumental in the success of applications that were conducted in real time being reliable and fast.

Among the most notable computers using magnetic core memory were those contained in the command module (CM) and lunar module (LM) of each *Apollo* flight to the Moon during the late 1960s and early 1970s. The shoebox-sized *Apollo* guidance computer (AGC), developed by the Instrumentation Laboratory at MIT, used a magnetic core memory with a RAM of just 4 K (words) and ran at a speed of 2 MHz – a specification exceeded by many of today's electronic toys for children (Figures 1.12 and 1.13).

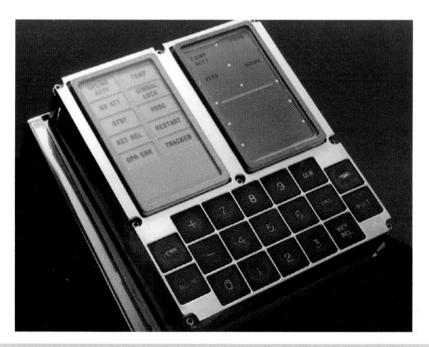

Figure 1.12. The *Apollo* guidance computer's user interface (NASA).

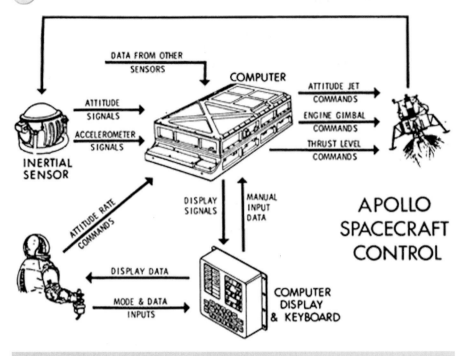

Figure 1.13. The *Apollo* LM primary guidance and navigation system (PGNS), crucial to a successful landing on the Moon's surface (NASA).

During the first Moon landing – that of *Apollo 11* on July 20, 1969 – the memory on board LM *Eagle's* computer became overloaded with data from the rendezvous radar, which had erroneously been switched on prior to the descent to the lunar surface. The error caused a confused computer to plot the descent of Neil Armstrong and Buzz Aldrin toward a landing amid a field full of large boulders – not the most desirable environment in which to set the fragile craft down. Armstrong had no option but to take manual control of the descent, and with just 30 s of fuel remaining he landed the craft beyond the danger zone on some smooth ground in Mare Tranquillitatis, a site now forever remembered as Statio Tranquillitatis (Tranquillity Base).

Integrated Advances

Hard on the heels of the invention of magnetic core memory, another major technological leap forward came in the form of the integrated circuit, invented (independently) in 1958 by Jack Kilby of Texas Instruments and Robert Noyce of Fairchild Semiconductor. Integrated circuits miniaturized and minimized electrical circuitry, increasing their effectiveness and decreasing their production cost. Instead of hefty hand-assembled electrical circuits containing masses of painstakingly interconnected wired components, integrated circuits are mass produced by

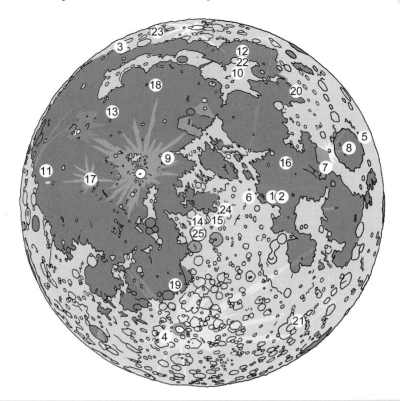

Figure 1.14. Map of the Moon showing the location of craters named in honor of people mentioned in this chapter. Key: 1 – Aldrin; 2 – Armstrong; 3 – Babbage; 4 – Brown; 5 – Cannon; 6 – d'Arrest; 7 – da Vinci; 8 – Eckert; 9 – Eratosthenes; 10 – Eudoxus; 11 – Galilei; 12 – Galle; 13 – Herschel (Caroline); 14 – Herschel (William); 15 – Hipparchus; 16 – Jansen; 17 – Kepler; 18 – Le Verrier; 19 – Lippershey; 20 – Maury; 21 – Metius; 22 – Mitchell; 23 – Pascal; 24 – Pickering; 25 – Ptolemaeus (photo by Peter Grego).

photolithography onto thin slices of semiconductor material (chiefly silicon, but also germanium and gallium arsenide). Through a series of photolithographic processes, integrated circuits can be built up to contain a vast number of transistors, resistors, and capacitors, all made from the same semiconductor material and arrayed in microscopic detail. Mass production rapidly brought production costs down, and the person on the street began to see and feel the rolling, ever-growing impact of the computer revolution. Before long, everyone was aware that silicon chips were an integral part of a computer's brain.

Noyce and his colleague Gordon Moore eventually left Fairchild Semiconductor to found a new company, Intel. There they developed a solid-state RAM circuit on a silicon chip; although this only had a 64-bit capacity, it was the size of a single 1-bit ferrite ring in magnetic core memories. During the early 1970s, Intel's RAM chips quickly began to replace the old-style magnetic core memory.

Moore's Law

In 1965, Moore observed that there was an exponential increase in the number of transistors that could be placed on an integrated circuit, amounting to a doubling of processing power every 18 months or so. During the 1970s the number of transistors on a single chip leapt from 10,000 to 50,000, passed the 100,000 mark in 1982, reached the 10 million mark by 1995, and hit a staggering 2 billion with the launch of Intel's Itanium Tukwila processors in 2008. Known as 'Moore's law,' this trend has continued for half a century and is liable to continue for at least another decade until the limits of miniaturization are reached at atomic levels – and some think even this may not represent the boundary of miniaturization. Moore's law (firmly established as a main goal of the computing industry) also holds true for other aspects of computing, such as RAM capacity, processing speed, and even the resolution of CCD chips in digital imaging devices. In accordance with Moore's law, personal computers have, year after year, increased in their processing speed, the amount of onboard RAM, and hard drive storage space, and the sophistication and quality of peripherals and hardware have improved beyond measure.

CHAPTER TWO

Computers Get Personal

Computers are pretty much everywhere these days, and it's fairly safe to say that anyone born after 1980 in a developed country won't remember a time when they weren't around. But it wasn't so very long ago that the idea of having a computer in the home for personal use (perhaps pleasure, even) was pure science fiction.

One very early personal computer was a sleek-looking, futuristically styled machine with the equally futuristic name of ZX81, manufactured by the pioneering British company of Sinclair Research (Figure 2.1). Timex Sinclair produced a clone of the ZX81 for the US market under the name of TS1000. The bare-bones ZX81 boasted a full kilobyte of RAM with which to perform its computational duties – that's just 1 kb, or around a millionth the size of the 1 GB RAM of most current home computers! The output, which required your own TV set to display, was monochrome, and the graphic display measured 32 characters wide by 24 high. Coarse graphics were possible using regular alphanumeric characters (black-on-white and white-on-black) and a set of 21 special graphics characters. Using this venerable device it was fairly easy (even for a not-particularly-mathematically-minded schoolboy) to write simple computer programs for the ZX81 in a language known as BASIC. This author eagerly applied that computer for astronomical use, using it to draw his own (admittedly, exceedingly simple) constellation maps and charts of lunar craters (Figures 2.2 and 2.3). He even found it possible to set in motion a miniature Solar System using moving characters as planets and comets, each of which had its own orbital velocity and variation in speed according to its distance

P. Grego, *Astronomical Cybersketching*, Patrick Moore's Practical Astronomy Series, DOI 10.1007/978-0-387-85351-2_2, © Springer Science+Business Media, LLC 2009

from the Sun. It was not exactly a snap to run these programs; they were stored on cassette tape and fed into the computer in all their screeching, ear-tingling glory, via an audio cable whenever they were required. Nevertheless, riding the crest of the first computer wave to hit the masses was a truly exciting experience.

Figure 2.1. The ZX81 had a small but fairly versatile set of characters.

Soon afterward I found myself pushing the burgeoning technology to its graphic limits (or so I like to think) as I experimented with computer art on a Commodore 64 as part of an art foundation course at college. It's odd to think now that the computer mouse was not part of our equipment – such an item we nowadays consider essential hadn't really come into general use by 1984. All the input had to be performed using the keyboard alone, which made the process pretty long and laborious. Yet regardless of this hurdle, the finished results seemed to be amazing – it really did appear possible to emulate, on-screen, such things as the colorful geometric artwork of the great Piet Mondrian (1872–1944) using this newfangled technology. Printing it out, however, was a different story – alas, at that time there was not even such a thing as a generally available color printer for the average PC user.

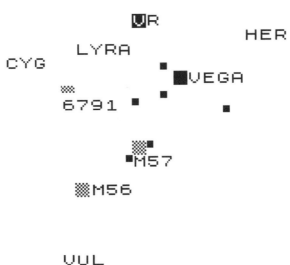

Figure 2.2. One of the author's earliest attempts at cybersketching. The author spent many hours compiling his own atlas of the constellations using the limited capacity of the ZX81. He has long since lost the tape upon which the atlas was stored, but this recreated graphic of the constellation of Lyra shows the kind of sophistication (!) capable on the machine (photo by Peter Grego).

Internationally renowned UK-based space artist David A Hardy has always experimented with ways to enhance his magnificent visions of the cosmos, including the use of photographic techniques; he had his own darkroom where he did his own developing and printing, derivative work, high-contrast effects with Kodalith, and so on. When computers arrived on the scene, Hardy quickly saw their potential. In 1986 he invested in an Atari 520 (512 k RAM), later progressing to a 1040 and eventually an Atari Falcon. Hardy used the digital painting program *Degas* which had 256 colors but interpolated them to look like 512; of course, one downside of such low-resolution graphics was that the large pixel size gave everything a somewhat blocky and jagged edge. He also used a program called *Art Director*. Even with the limitations of these early graphics programs, Hardy managed to use this setup to produce some interior illustrations for the American science fiction/fact magazine *Analog* and did the background graphics for the highly acclaimed computer game *Kristal* for Atari and Amiga (Figures 2.4 and 2.5). It's worth noting that these were the days before graphics tablets and the creative freedom delivered by the digital stylus; all the input for this artwork was done using only keyboard commands and a two-button mouse.

LUNAR SEAS

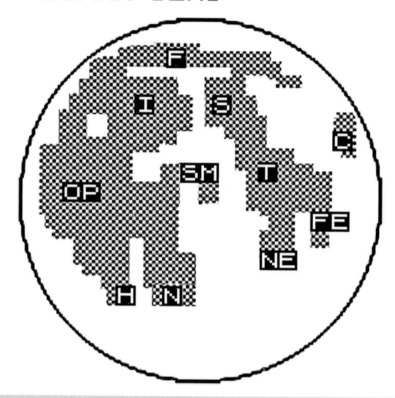

Figure 2.3. Another recreation of one of my early attempts at cybersketching – a Moon chart prepared on a ZX81 computer showing the location of the main lunar seas (photo by Peter Grego).

The Kristal contains some magnificent graphics

February 1989 Atari ST User 109

Figure 2.4. A volcanic alien world, created on *Degas* by David A. Hardy for the video game *Kristal*. The painting was created using just a mouse and keyboard (David A. Hardy).

Figure 2.5. An inhabited alien world, created on *Degas* by David A. Hardy for the video game *Kristal*. (David A. Hardy).

Apple Ladles on the Sauce

Computer technology – in terms of materials, hardware, and software – advanced in great leaps and bounds during the 1980s and 1990s. A major landmark was reached in 1984 when the US company Apple released their first Macintosh personal computer, a compact machine designed with ease of use in mind and aimed at the general consumer market (Figure 2.6). Housed within a portable (though hardly 'lightweight') beige cube, the Macintosh had a 9-in., 512×342 pixel monochrome display, beneath which lay a 3.5-in. floppy disk drive; it boasted 128 kb of RAM with which to perform its operations, but lacked a hard drive. Unlike other personal computers on the market at that time, the Macintosh featured a mouse and a graphical user interface (GUI) – a set of icons displayed on the screen in a virtual desktop which, when clicked upon, displayed programs within rectangular, overlapping, resizable windows. Menus within programs could be pulled down to execute a variety of commands.

To those previously restricted to a keyboard, the original Macintosh was a revelation. My own first encounter with this legendary piece of hardware took place in 1988 at the University of Aston in Birmingham, under the guidance of Professor John Penny, editor of the Birmingham Astronomical Society *Newsletter*. I soon found my feet and began to write astronomy articles on *MacWrite* and produced illustrations in a graphics program called *MacPaint*, using the mouse to manipulate numerous pen, paint, shade, and fill and shape effects (Figures 2.7 and 2.8). Although the graphics were pretty basic (by today's standards, at least) it seemed obvious even then that computers had great potential to be used as an aid to observing and recording astronomical subjects. My own first electronic observational drawings consisted of a series of Mars observations made during the planet's 1988 apparition; although the shading was created using a spray brush effect, creating black dots on a white background, they were reasonable efforts (providing the viewer look at them through narrowed eyes from across the room) (Figures 2.9 and 2.10).

Figure 2.6. A vintage Macintosh computer.

Apple's share in the personal computing market grew during the 1980s as improved versions of the Macintosh were introduced, complete with hard drives, increased RAM, and expansion slots. Also, a range of peripherals such as printers and scanners appeared. Among new software, all of which was produced by Apple itself, were spreadsheet and presentation software, and improved 'Pro' graphics and DTP applications. In 1989 Apple even launched a laptop, the Macintosh Portable, and the company later pioneered laptop ergonomics with the Powerbook 100, placing the keyboard behind a wrist rest and using a track-ball pointing device to replace the mouse, which was placed centrally at the base of the keyboard.

Competition with Apple's range of Macintosh computers rapidly gained ground in the form of PCs using MS-DOS/Windows operating systems. Microsoft Windows 3.0, released in 1990, was the first operating system to go head to head with the Macintosh operating system, offering a seemingly comparable level of performance and set of features, and presented a virtual desktop GUI. Importantly, PCs running Windows 3.0 were less expensive than their Apple rivals. Painfully aware of the new competition, Apple sued their rivals Microsoft, with the accusation that Apple's copyrighted GUI was being infringed. In addition to losing their case after a 4-year period of litigation, Apple lost not only the support of many

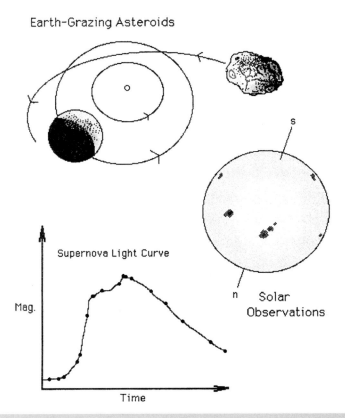

Figure 2.7. Author's original Macintosh illustrations from the cover of the BAS *Newsletter*, December 1987 (photo by Peter Grego).

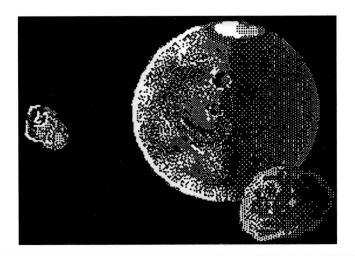

Figure 2.8. Author's original Macintosh illustration from the cover of the BAS *Newsletter*, October 1988, showing Mars with Phobos and Deimos (photo by Peter Grego).

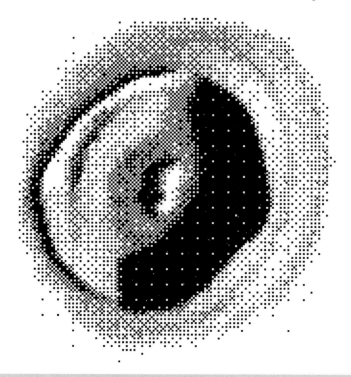

Figure 2.9. Early example of a lunar drawing, drawn (using a mouse) and printed on an Apple Macintosh, based upon the author's observation of the lunar crater Lansberg made on May 8, 1987 (photo by Peter Grego).

within the industry but many consumers too; by mounting a legal challenge to Windows, the company was accused of stifling development and choice by creating a monopoly on a particular style of GUI.

Apple eventually replaced the name 'Macintosh' with the snappier, friendlier 'Mac' and prided itself on the sleek design of its products. A major target area for its goods was the graphics and professional user, both personal and corporate. Programs such as the desktop publishing classic *Quark XPress* and the graphics programs *Photoshop* and *Illustrator* (both by Adobe) were among the cornerstones bolstering the Mac's reputation as being the most desirable graphics and DTP computer. In 1998 Apple produced the eye-catching iMac, an all-in-one device with monitor and computer encapsulated within a streamlined, translucent case (originally a cool blue, but later available in other colors). Apple dispensed with the usual large and unsightly connections (SCSI and ADB) and instead featured two USB ports for peripheral connections. Phenomenal sales of the iMac helped revive the flagging company.

Many Mac users today are professional writers and graphics people. Space artist David A Hardy obtained his first PowerMac 7100 (with 512 MB of RAM) in 1991; since then he has gradually moved up. He now has a 24-in. iMac (with an Intel Core Duo chip and 4 GB of RAM) running OS X 10.4.11 'Tiger.' Hardy's most used program – and one that he admits not being able to do without, for digital graphics

MARS SHRINKS AND FADES AT THE END OF A SPECTACULAR APPARITION

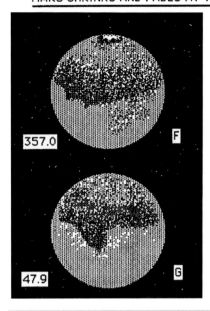

25th November 1988
21h 18m - 21h 26m UT
4" Wray refractor, Bearwood ×128

Not much detail was discernible
on Mars. Note how the south cap
has shrunk in size.

28th December 1988
20h 41m - 21h 56m UT
12" f.5 Dorridge, × 150

Even using a 12" reflector, Mars
is now a difficult object upon which
to discern details. The south polar cap
was not visible.

Figure 2.10. Author's observations of Mars from 1988, drawn on a Macintosh (photo by Peter Grego).

at least – is Adobe *Photoshop CS2*. He doesn't use any 3D programs except *Terragen* to generate landscapes (but he does produce his own bump maps in *Photoshop*) and *Poser 7* for figures.

The latest Apple iMac is contained within a stylish aluminum carapace; Apple also produces a Mac mini desktop model, the MacBook, the incredibly slender MacBook Air (the world's thinnest laptop), and MacBook Pro laptops. Although Mac OS X 10.5 'Leopard' is the latest Mac operating system, Macs are now able to run other operating systems such as Linux and Microsoft Windows.

The Windows of Change

During the 1990s, Microsoft Windows became the world's most popular operating system (in terms of the sheer number of its users). Windows went through a variety of incarnations, including Windows 3.1 and Windows NT (both introduced in 1992), Windows 95 (1995), Windows 98 (1998), Windows 2000 (2000), Windows ME (Millennium Edition, 2000), Windows XP (2001), and Windows Vista (2007). These operating systems were not without their faults, and critics of Microsoft were (and still are) vociferous. System crashes with the earlier operating systems were common, and they were all prone to malicious attacks by such diabolical electronic entities as worms, Trojans, and viruses. Mac users, meanwhile, experienced few of these problems.

Windows XP (Pro version) remains the most stable operating system that this author has used to date; the OS seized up only a handful of times over a 4-year time span, during which it was used on a daily basis, largely for writing, editing, and illustrating. However, an upgrade to a new machine running Windows Vista caused a great deal of frustration. A number of programs that had worked without a hitch in XP now refused to work on the new OS, even using Vista's backward-compatible mode. Other programs that initially appeared to work perfectly well went on to malfunction in certain areas. For example, Corel PhotoPaint 8 and PaintShopPro 8, both of which had performed sterling service in XP, showed a number of annoying glitches. Additionally, various pieces of computer hardware that had been installed in XP were required to be reinstalled in Vista. Unfortunately, Vista-compatible drivers are difficult to find for many peripherals produced before Vista came on the market. One example that caused great annoyance involved a perfectly good Philips ToUcam Pro PCVC740K, a webcam that had been used to capture hundreds of high-resolution lunar and planetary images. Sadly, no Vista-compatible drivers are available for this camera, a situation which rendered a perfectly functional device unusable.

Desktop Essentials

In the late 1990s only one in three UK and US households had a personal computer; that figure had more than doubled for both countries a decade later. However, the number of homes with a PDA is far lower, at around the one in six mark. Because of the PDA's mobile business applications the percentage of PDA ownership is greater in professional, higher income households (PDAs replaced Filofaxes during the 1990s). In this book we hope to demonstrate that computers can be valuable astronomical tools and that laptops, tablet PCs, UMPCs, and PDAs can be versatile friends to the everyday observational amateur astronomer – not just a flashy high-tech status symbol owned by the well-off.

It's probably fair to say that most people in the Western world can afford a personal computer. Because of Moore's law (see earlier), computer specifications are improving continually. What may seem like a white-hot piece of equipment today will probably seem like a damp squib (to the computer aficionado, at least) in a couple of years' time.

Specs Appeal

Although there appears to be an abundance of choice in the personal computer market, making the right choice can be a rather complicated process. Your ideal computer's specifications depend on your computing needs (both current and anticipated), and the decision is ultimately tempered by your budget. Graphics work with image manipulation programs such as Adobe *Photoshop CS2* requires a fairly fast computer running at least Microsoft Windows 2000 (SP4) or Windows XP (SP1 or SP2), with a minimum of 384 MB of RAM, a monitor capable of 1024 × 768 pixel resolution, and 650 MB of available hard disk space.

To make it a little easier to assess the efficiency of your own computer's hardware setup, and to choose the right sort of software for it, Windows Vista has introduced a Windows Experience Index. This built-in application measures a computer's hardware capabilities and software configuration and allocates a base score number for it based on the lowest subscore attained by any component of the computer's hardware. The scale currently ranges between 1 and 5.9; the higher the base score, the better your computer's performance based on the measured criteria, especially when more advanced and resource-intensive tasks are required.

A base score ranging between 1 and 1.9 is the minimum specification required to run Vista and deliver basic performance. A base score of 3–3.9 will enable Windows Aero (a supposed 'next generation' desktop interface) to be utilized; graphics-intensive programs and games can be run on such a machine. Computers with ratings from 4 to 4.9 allow high-definition video and the use of high-resolution monitors, while the highest-end computers are rated between 5 and 5.9 and are of very high performance.

As of the time of writing (July 2008) a year-old machine with Vista Home Premium OS (32 bit) uses a Pentium Core 2 Duo processor and has 1014 MB of RAM and two internal hard drives with a combined capacity of 500 GB. It is perfectly adequate for writing, desktop publishing, graphics, and image manipulation needs, although with a Windows Experience Index base score of 3 it might benefit from some beefing up by adding more RAM and a slicker graphics card.

Consider the Aged

Of course, it's perfectly possible to run older versions of well-known astronomical programs such as *RedShift* and *Starry Night* and experience the graphics capabilities of older versions of good drawing/image manipulation programs such as Corel's *PhotoPaint* and Adobe's *PhotoShop* using, say, an early 2000s machine running Windows 98 SE. Such a machine will also be able to link with a PDA through a cable using the Windows *ActiveSync* program.

Perfectly good used computers – with monitor, keyboard, and mouse thrown in – are available from just about every backstreet emporium for less than the cost of a decent pair of jeans. Sure, a used computer of a few years' vintage may not be able to run the latest Windows OS, and it may chug along at a blisteringly leisurely pace accompanied by a strange assortment of internal clicks and rattles, but it will probably perform satisfactorily as long as the user appreciates its many limitations. Chief among these limitations are the vintage computer's relatively slow processing speed, its modest onboard RAM and hard drive capacity, its outmoded OS, and inability to run the latest programs, which are written for much higher-spec computers. A vintage computer's physical limitations include its ergonomic footprint (older computers can take up quite a large space), its lack of adequate ports and connections, and the structure of its motherboard, which may not be friendly to modern upgrades of its memory and CPU.

Given the increasing sophistication of computer programs and the ever-growing demands that they make on the computer's processing power, there comes a time when every computer user considers upgrading the computer's components to make it a more capable machine. Given the alternative of buying a brand new computer, upgrading the OS software, RAM, and hard drive memory can, in many cases, be an attractive and cost-effective course of action. The most common upgrades to a computer's physical contents include expanding its memory capacity by replacing its RAM cards, replacing or adding an additional hard disk drive, replacing the CPU, and even replacing the entire motherboard.

Bettering RAM

Motherboards inside computers are usually provided with a specific type of slot into which RAM memory is inserted, and upgrading RAM may appear to be a simple task. However, because of the large number of types of RAM and their differing physical sizes and shapes, upgrading can prove to be a tricky business (Figure 2.11). Common types of RAM for desktop computers include

SIMM (Single in-line memory module) 30-pin: Used in vintage desktop computers with Intel 286, 386, and 486 processors from around 1982 to the mid-1990s;

SIMM 72-pin: Largely used in old desktop computers with Intel 486 and Pentium processors, from around 1993 to 1997;

SDRAM (Synchronous dynamic random access memory) 168-pin: Commonly found in older computers from around 1993 to 2007;

DDR (Double data rate) 184-pin DDR2 240-pin DIMM (Dual in-line memory module): Used in newer computers;

DDR3 240-pin DIMM: Used in some of the latest desktop computers.

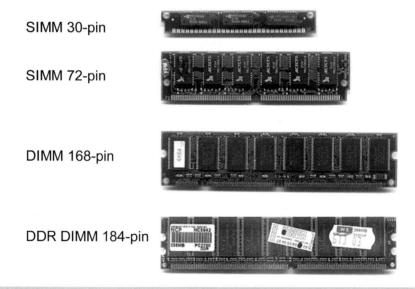

SIMM 30-pin

SIMM 72-pin

DIMM 168-pin

DDR DIMM 184-pin

Figure 2.11. Various types of RAM: SIMM 30-pin, SIMM 72-pin, DIMM 168-pin, DDR DIMM 184-pin (Wikimedia Commons).

Going for a Drive

Virtually all modern desktop computers are equipped with one or more hard drives upon which the operating system and other programs are stored. In most personal computers the main HDD is the 'C drive,' since Microsoft Windows designates the letter C to the primary partition on the main hard drive by default.

Hard disk drives are termed non-volatile storage devices, which means that they retain all the data stored upon them even when they are switched off. Hard drive data are encoded upon a stack of rapidly rotating platters with magnetic surfaces; these data are read and written on by means of read–write heads attached to actuator arms that scan the surface of the platters. IBM developed the first recognizably modern hard drive more than 50 years ago – an electromechanical beast that consisted of no fewer than fifty 60 cm (24-in.) platters and that had a capacity of 5 MB. In 1981 the Seagate company introduced the first hard drive for personal computers, a 5.25-in. model with a 40 MB capacity and data transfer rate of 625 Kbps.

Modern hard drives come in two form factors – the 5.25-in. and 3.5-in. variety. They are hermetically sealed units containing several stacked platters spaced a few millimeters apart, to make room for the read/write heads on actuator arms (Figure 2.12). During disk operation, the air flow generated by the rapidly spinning (typically 7,200 rpm in a desktop computer) platters causes the read/write heads to fly just a few hundredths of a millimeter above the surface of the platters – closer than the width of a smoke particle. Data are stored on each platter's surface in sectors (concentric circles) and tracks (pie-shaped wedges). Most of the data aren't stored in any particular order; the HDD makes use of any convenient empty space (such as that freed by deleted files), so that after a while the data get 'fragmented,' scattered around the hard drive. This gradually slows down the process of accessing data stored on the hard drive. Defragmenting the HDD regularly (see below) will enable things to run as smoothly and as quickly as possible.

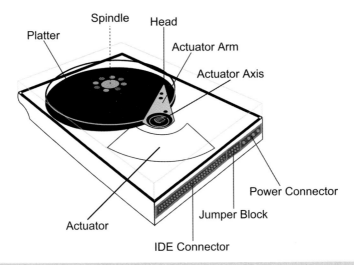

Figure 2.12. The main features inside a hard disk drive (Wikimedia Commons/Peter Grego).

Hard drives are so finely constructed that a microscopic particle of dirt in the works might easily cause a head to crash into its platter, scouring off its magnetic data and causing the drive to fail catastrophically. Unless a user has a very healthy bank account and the data are considered absolutely essential to retrieve, hard drives that have crashed through hardware failure are usually thrown away, their data never to flicker across the user's monitor again. There are other causes of failure – including software malfunction or virus, for example – that may not render the data permanently inaccessible. However, hard drives are mechanical devices subject to wear and tear, and each one will eventually succumb to the ravages of old age. To ensure optimum HDD health, the following steps are recommended to all computer users:

- Back up all the data you consider to be vital – including files, texts, and images – on another HDD or on removable media.
- Monitor the health of your hard drive (and that of your computer system) by using diagnostic programs. Some of these programs are provided with your computer's operating system, and some are available as downloadable freeware or commercial programs. Diagnostic software is capable of identifying and predicting problems, as well as offering solutions. (Self- Monitoring, Analysis and Reporting Technology, SMART, is built into all new hard drives.)
- Use system tools such as the defragmenter and clean up temporary files stored in the cache (such as temporary Internet files) with Disk Cleanup. Do this on a regular basis, at least once a month if you're a moderate computer user.
- Make sure that your antivirus software is kept fully up to date. Viruses have been known to modify and completely delete the contents of hard drives; in some cases even a complete drive reformatting isn't enough to remove the virus. Prevention is infinitely better than cure in the world of computing.
- Keep your computer operating system's software up to date by downloading the latest patches, service packs, and updates from your OS provider's own web site.
- Completely remove any programs that you don't use and delete any unwanted images and files; these all take up valuable HDD space.

Since your computer's operating system, along with all of your programs and saved files, is stored on its hard drive, replacing the hard drive with a faster one with larger capacity can seem a little daunting. Replacing a hard drive with a new one requires the installation of an operating system (whether it's the OS you're used to or a newer version) and a reinstallation of your programs. Many choose to simply add another hard drive rather than replace the existing one, which often entails just opening up the computer case and plugging it into a spare power source and connecting it to an IDE cable.

The Mouse Squeaks Up

Today, the mouse is as familiar a piece of computer hardware as the keyboard and monitor. As we have seen, most personal computers of the early 1980s didn't have the luxury of a mouse, and it was only the eye-opening arrival of the GUI (graphical user interface) in the Apple Macintosh and machines running the first versions of Microsoft Windows that the mouse became as important an input device as the keyboard. However, the mouse was invented way back in 1963 by Douglas Englebart at the Stanford Research Institute, and the earliest reference to such a device as a 'mouse' was published in *Computer-Aided Display Control* (1965) by William English, Englebart's colleague. The original mouse had two external wheels whose rotations were read by a mechanical encoder to determine its movements relative to the horizontal plane beneath the mouse. English went on to invent the ball mouse in 1972, which used the movements of a small hard rubber ball and internal rollers (one horizontal, the other vertical) in contact with it to mechanically encode the mouse's movements and translate it to move the on-screen cursor.

Further developments at the École Polytechnique Fédérale de Lausanne (Lausanne Federal Polytechnic School) in Switzerland gave rise to the appearance of the first modern computer mice. Mechanical encoders were replaced by more accurate optical encoders, and the design was improved to feature three buttons on the mouse's upper leading edge, which could be operated by the fingertips. In 1981 the Swiss company Logitech was the first company to produce mice, and derivatives of these so-called opto-mechanical mice were most common during the 1980s and 1990s (Figure 2.13).

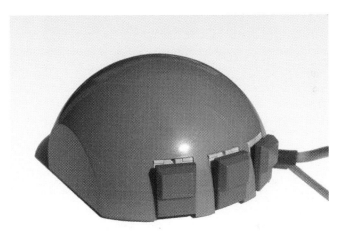

Figure 2.13. A mouse of the early 1980s, designed by André Guignard and retailed as Logitech's first mouse (Stéphane Magnenat, Wikimedia Commons).

Some mice attempt to improve pointing precision by having a trackball on their upper surface, including devices with large trackballs to assist children or adult users with impaired motor skills. Among the downsides of trackballs and opto-mechanical mice is their annoying ability to gather a residue of dirt that coats the surface of the ball and the internal rollers, reducing their accuracy and performance over time, so they require regular cleaning and maintenance. Opto-mechanical mice were eventually outclassed and outsold during the late 1990s when the optical mouse became widely available. Instead of encoding the movements of a physical ball on an underlying surface, the optical mouse uses a light-emitting diode (LED), which projects a beam onto the work surface; the reflection of this beam is constantly monitored by photodiodes and processed by image processing chips inside the mouse to detect movement. Optical mice are much more accurate than their balled counterparts and require less maintenance in terms of cleaning.

Although most mice fit beneath the palm of an average adult-sized hand, many different mouse shapes have been produced over the years, some of them designed ergonomically to nestle more comfortably in the palm and suiting either right- or left-handed people. One of the most useful developments came with the addition of a central scrolling wheel between the main mouse buttons to facilitate rapid movement up and down the display or to adjust the display's zoom level.

For years, mouse users had no option but to contend with the mouse's lead that connected it to the computer through a serial, PS/2, or USB port; leads have an annoying habit of occasionally getting in the way of things, tangling up, snagging, and accumulating dirt. Thankfully, mice were liberated when their tails were chopped off. During the 1990s there appeared the cordless mouse, which communicated with the computer via its own USB RF (radio frequency) receiver. The same technology also saw the appearance of other cordless RF devices, notably the cordless keyboard. Cordless mice, however, are hungry for power, and it is not uncommon for them to require fresh batteries (usually AA- or AAA-sized cells) every month or two. The best solution is to use rechargeable batteries in the cordless mouse, or better still, to save having to physically take out the batteries to recharge them, a mouse that recharges itself via its own desktop cradle, which doubles as a receiver for the mouse's RF signals (Figure 2.14).

Figure 2.14. The author's main mouse, a Packard Bell PB-EWRO1 cordless rechargeable optical mouse, shown in its cradle (photo by Peter Grego).

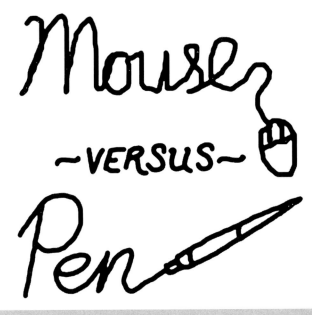

Figure 2.15. Simple graphic drawn with a mouse ('Mouse') and with a graphics stylus ('versus Pen'), demonstrating the superior capabilities of the latter in producing more accurate graphics (photo by Peter Grego).

One of the most significant recent advances in mouse technology came in the late 1990s with the first laser mouse; the technology was generally available for the first time in 2004 when Logitech introduced its MX1000 Laser Cordless Mouse, and various other companies have since followed suit and introduced their own laser mice. Laser mice work in essentially the same way as using optical mice, except that they use an infrared laser diode instead of an LED to illuminate the surface beneath them. The use of lasers allows for better resolution of the illuminated surface, increasing the mouse's tracking power and accuracy.

Mice are used extensively to operate computer graphics programs and to edit, manipulate, and enhance images and graphics. However, their effectiveness in controlling the on-screen cursor with sufficient accuracy to produce the fine details of a drawing is limited by ergonomics. Most humans have enough difficulty in producing accurate drawings with a regular pencil and paper, let alone using an ostensibly counterintuitive device like a mouse to produce drawings on a screen.

Graphics Tablets

The notion that a stylus could be used to input handwritten data into a machine was recognized around long before the invention of electronic computers. In 1914 Hyman Goldberg filed a US patent for a 'controller' that allowed recognition of handwritten numerals to control a machine in real time, and in 1938 George Hansel filed a US patent for machine recognition of handwriting. Glimmers of a real breakthrough came in 1957 when T L Dimond of Bell Laboratories developed the idea of an electronic 'Stylator' system that used a computer to analyze the stroke sequence and direction of characters written onto an energized plastic tablet and converting this information into characters by comparing it with a built-in dictionary.

In 1964 the RAND (Research ANd Development) Corporation produced the first modern graphics tablet, the Grafacon ('Graphic Converter'), which encoded the horizontal and vertical coordinates of a magnetic signal produced in a grid of wires beneath the pad's surface via a stylus. Dreams of character recognition finally became a reality in the early 1980s, when Charles Elbaum (Professor of Physics at Brown University) cofounded the company Nestor and developed the Nestor-Writer handwriting recognition system.

In the early days, stylus input onto a graphics tablet promised to be quicker, easier, and more intuitive than using a keyboard to control computer programs; with roughly the same size and shape as a normal pen, the stylus could be used to perform all the same functions as a mouse. Koala Technologies produced the first popular graphics tablet, the Koala Pad, in 1984, which was intended to be used with a number of early home computers such as the Commodore 64 and Radio Shack's TRS-80 for basic drawing and painting. More sophisticated graphics tablets began to appear as PC peripheral devices during the early 1990s, with companies such as Wacom producing a range of devices in a variety of sizes for use with Microsoft Windows and Mac OS computers.

The Wacom Graphire, along with many other graphics tablets in the product range of Wacom and other companies, is said to be a passive tablet. Passive tablets get what little power they require directly through the serial or USB port to which they are connected, and the tablet, mouse, or stylus do not require batteries or a separate power supply. Passive tablets make use of electromagnetic induction, where a horizontal and vertical matrix of wires beneath the tablet's surface generate an electromagnetic signal received by the stylus' LC circuit and then in turn are read by the tablet's wires in receiving mode. In this way, how the stylus tip coordinates with respect to the tablet's surface can be gauged with great accuracy. Most graphics tablets are also capable of determining the amount of pressure being applied by the tip of the stylus onto the tablet; older (or cheaper) tablets have a sensitivity of 256 levels, a standard graphics tablet is sensitive to 512 pressure levels, and a high-end professional tablet has a sensitivity to 1,024 levels of pressure.

The venerable Wacom Graphire of 2000 vintage is connected to the computer via an ordinary serial interface and has a cordless mouse in addition to a stylus (this works along the same principles as the stylus). Of the same approximate dimensions as a mouse mat, the actual writing area on the tablet measures 90 × 130 mm. The mouse, which tends to be on the sluggish side and is not as accurate as a regular optical mouse, still occasionally comes in handy when the optical mouse needs recharging. The stylus has a 512-level pressure-sensitive writing tip and a handy pressure-sensitive eraser at its other end; a two-function rocker switch on the mid-shaft of the stylus operates the left and right mouse button functions, but the buttons can be programmed to operate virtually any function. In comparison with the latest graphics gear, this tablet is admittedly not in the same league as the best modern ones, but it can serve many purposes very well (Figure 2.16).

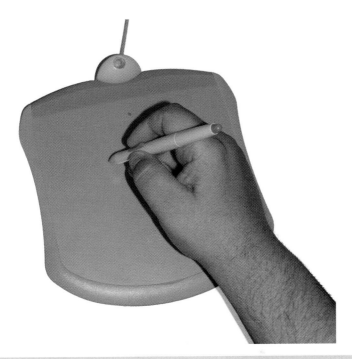

Figure 2.16. Vintage but valued, the author's Wacom Graphire graphics tablet, used since 2000 and still going strong (photo by Peter Grego).

Active tablets make use of a powered stylus, which transmits a signal to the tablet. Since active styluses require batteries, they are usually somewhat heavier than their passive stylus counterparts. However, active tablets have no need to transmit signals to the stylus, only to receive them, so they are considered to be less prone to jitter and therefore more accurate.

Figure 2.17. The active stylus of the Hammerhead HH3 tough tablet PC, showing the AAAA battery and the internal electronics (photo by Peter Grego).

Digital Notepads

Digital notepads are portable graphics tablets that convert regular handwriting and drawings made on overlaid sheets of normal paper into a digital format (Figure 2.18). Digital notepads differ from regular graphics tablets in that they are portable (but permanently connected to a PC) and are capable of storing graphics in their built-in memory or onto memory cards. They also tend to be more sensitive than graphics tablets in that many of them are capable of reading input through a centimeter or more thickness of paper. This means that a regular notepad or drawing pad may be used in conjunction with them.

Figure 2.18. A Digimemo A402 digital notepad (showing a simulated astronomical sketch) (photo by Peter Grego).

The input device, a ballpoint pen, has a battery-powered built-in electromagnetic transmitter whose exact position is registered by the tablet when pressure is applied to the ballpoint. Inbuilt storage varies from device to device; the Digimemo A402 can store up to 50 pages of writing and drawing, while the Digimemo A501 has 32 MB flash memory and can store up to 160 completely filled A5-sized pages. Drawings and/or text are downloaded from the device onto a computer through a USB connection or by memory card; handwriting can be converted to text using recognition software, and drawings can be converted to standard image files.

Digital notepads aren't intended to produce high-resolution works of art, but they can be useful for basic cybersketching and astronomical note taking in the field.

Getting the Picture

Video displays are, of course, an absolutely indispensable component of modern computers. It wasn't so very long ago that many personal computers – the Sinclair ZX81 and Spectrum, the Atari 400 and 800, the Commodore VIC-20 and 64, to name but a few models of the early 1980s – required plugging into the aerial socket of a regular television set. Since then, computer monitors have developed into truly wonderful devices capable of displaying graphics in glorious detail. Like other cybersketchers and producers/collectors of astronomical graphics and images, it is possible to lose yourself for hours on end simply browsing the image contents of your computer's hard drive. A good monitor is essential if you want to enjoy viewing your work on-screen.

View Tubes

CRT (cathode ray tube) displays have been associated with computers for more than half a century, but the technology behind them was developed far earlier. The first cathode ray tubes were developed in the late nineteenth century by the English scientist William Crookes to prove the existence of 'cathode rays.' Crookes' tubes were evacuated cone-shaped glass objects with an electrical filament at the small end and a phosphor-coated surface at the other end. The 'cathode rays' appeared to be emitted by the filament, causing the phosphor screen to glow; if the path of the 'rays' happened to be intruded upon by something like a metal sheet inside the tube, then its shadow would be formed on the screen. It was later discovered that 'cathode rays' are, in fact, composed of a stream of electrons; the electrons energize the phosphor screen on impact, causing it to glow.

During the first half of the twentieth century scientists developed the idea to produce moving images on the phosphor screen using electron guns, focusing the electron beam and manipulating it by scanning the screen in much the same way as you're reading this book, but many thousands of times faster (even if you happen to be an accomplished speed reader), sweeping from left to right in successive rows from top to bottom. Thus, television was born, with the first regular broadcasts commencing in 1933 courtesy of the BBC (British Broadcasting Corporation) from the Crystal Palace near London.

Color television well and truly arrived in the 1960s. Color is produced when three electron beams strike a three-layered phosphor screen; each layer glows in a specific color, red, blue, or green. Different colors are created by mixing the three primary colors to varying extents; white is produced by a combination of all three beams, while black is produced in the beams' absence. The screen is also covered with a metal sheet containing a grid of minute holes, called a shadow mask, to produce a clearer image (Figure 2.19).

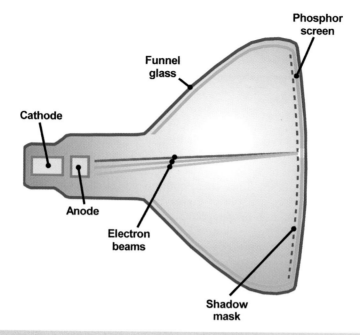

Figure 2.19. Simplified drawing of a color CRT monitor showing the main components (photo by Peter Grego).

CRT computer monitors use the same principles as old-style televisions, their electron guns being manipulated by the video data being fed to them from the computer's graphics card. In the early days, CRT monitors were monochrome, often displaying lurid green graphics. This author used one for a couple of years with an Olivetti PC-1 (1990 vintage) to write astronomical articles, and the monitor's verdant hue became less objectionable as time went by. In the 1980s an organization called VESA (Video Electronics Standards Association) was set up by a number of leading video adapter and monitor manufacturers in order to standardize the video protocols of color monitors. The main types are

Year	Standard	Resolution	Colors
1987	VGA (Video Graphics Array)	640 × 480	16
		320 × 200	256
1987	SVGA (Super Video Graphics Array)	800 × 600	256 to 16.7 million
1990	XGA (Extended Graphics Array)	1024 × 768	16.7 million
	SXGA (Super Extended Graphics Array)	1280 × 1024	16.7 million
	UXGA (Ultra XGA)	1600 × 1200	16.7 million
	WXGA (Wide XGA)	1366 × 768	16.7 million

It is possible to modify the resolution and number of colors displayed by a monitor within certain limitations, depending on the graphics capabilities of the monitor and the computer's video card (Figure 2.20). Some applications automatically adjust the display settings when they are run, while others require the display settings to be input each time the program is run; a feature in later versions of Windows allows the user to determine the display settings for each program.

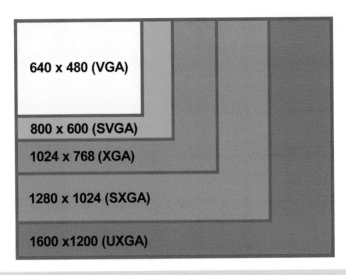

Figure 2.20. Resolution comparisons for four different 4:3 ratio screens (photo by Peter Grego).

CRT monitors are usually produced in sizes ranging between 15 and 22 in. (38–53 cm) across, measured diagonally across the screen. On all CRTs the actual area available for viewing is usually a couple of centimeters smaller than its stated size because the monitor's bezel hides a little of the CRT's edges. Like CRT televisions, the vast majority of CRT monitors have the standard 4:3 screen geometry (a size ratio of 4 units wide by 3 units high). Larger CRT monitors can be extremely heavy, and they require a pretty sturdy desk.

It's obvious even to the untrained eye of a novice computer user that all modern CRT monitors are not equal in terms of image quality. In addition to a monitor's physical size, resolution, and chromatic depth (the number of colors it is capable of displaying), several other factors influence image quality. Image quality is affected by the CRT monitor's type of shadow mask; shadow masks composed of small circular holes are capable of rendering the smoothest looking lines and curves and producing the sharpest detail, and they are consequently often the CRT of choice for those involved in graphics work. Another kind of shadow mask, the aperture grille, uses an array of straight vertical slots; these are most often found in flat screen CRTs and they deliver the best colors, produce the least distortion, and reduce the effects of external glare (Figure 2.21).

Another important factor determining image quality is a CRT monitor's dot pitch – the spacing between the screen's individual pixels (note that dot pitch is not applicable to CRTs with aperture grilles). Dot pitch is given in fractions of a millimeter, and the lower the figure, the clearer the image, since any given area will contain more pixels, and the image will retain its visual

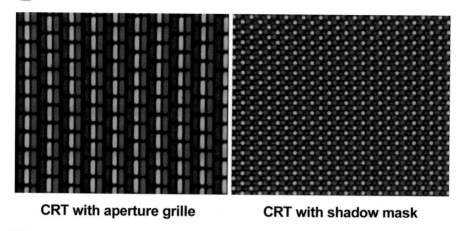

CRT with aperture grille   **CRT with shadow mask**

Figure 2.21. Close-up of aperture grille and shadow mask CRT screens (photo by Peter Grego).

integrity on increasingly closer viewing (Figure 2.22). Dot pitch is measured along the diagonal line between dots of the same color, but some brands have chosen to state the distance between horizontally adjacent dots as the dot pitch, ignoring the vertical spacing of the dots (Figure 2.23). This incorrect method of measuring dot pitch is misleading for the consumer because it results in a smaller figure, wrongly suggesting that the monitor has a better resolution than it actually has. Entry-level CRT monitors typically have a diagonal dot pitch of 0.28 mm or more, while this is reduced to 0.26 mm or smaller in good quality monitors. High-end CRT monitors may have a dot pitch less than one tenth of a millimeter.

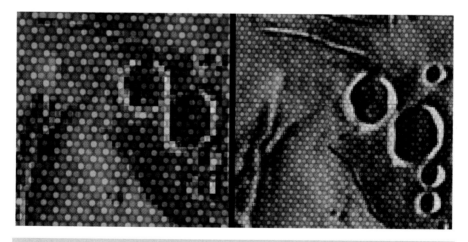

Figure 2.22. Simulation of a low-resolution CRT display (*high dot pitch*) compared with a higher resolution display (*low dot pitch*), both showing the same image and viewed from the same close distance (photo by Peter Grego).

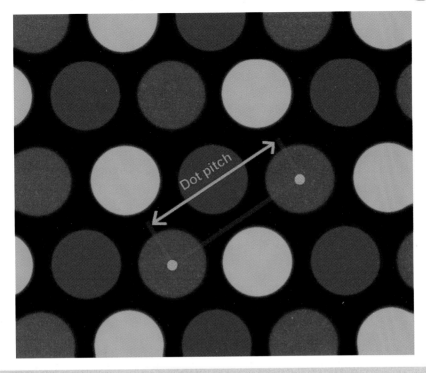

Figure 2.23. The correct method of determining dot pitch on a shadow mask CRT monitor (photo by Peter Grego).

Graphic applications require the sharpest possible images to be displayed by a monitor, and the quality improves with higher resolution in combination with a larger screen. A monitor with a minimum resolution of 1024 × 768 pixels and larger than 17 in. is best for graphics work, provided that the computer with which it is used has a video adapter card capable of handling the resolution size of your choice. In addition to a monitor's physical size and resolution, the rate at which the screen image is refreshed determines 'flicker'; the higher its refresh rate, the less flicker and consequently the more comfortable the image is to view. Eye strain can be caused by smaller monitors with refresh rates of below 75 Hz, while monitors larger than 19 in. may require refresh rates of at least 85 Hz.

Crystal Clarity

In the mid-2000s LCD (liquid crystal display) monitors overtook CRTs to become the most popular form of computer monitor. LCD monitors have a number of advantages over CRTs. They are far lighter in weight than CRTs and take up less desktop space because they are considerably thinner; they can be mounted on an adjustable wall bracket to free up more desk space, and they can be tilted to virtually any angle. LCD monitors have flat screens, are less prone to annoying reflections, and are more comfortable on the eye because they are flicker-free. LCD monitors typically use up about half the power of an equivalent-sized CRT, and they produce less heat.

One of the biggest disadvantages of LCDs in comparison with CRT monitors is their cost; small CRT monitors, for example, are currently about half the price of their LCD counterparts. It's also worth pointing out that the bottom has virtually fallen out of the used CRT market, and superbly graphics-capable CRTs that were once the envy of graphics professionals can be picked up for a real bargain. In terms of image quality, CRTs are generally thought to represent colors better than LCD displays and can handle a wider variety of resolutions. Finally, CRT screens are more durable and more capable of withstanding quite hard knocks than LCD monitors.

The apparently oxymoronically named liquid crystal display began to be developed in the 1970s. It was found that when an electric current is applied to a special kind of crystal called a nematic phase liquid crystal, its structure untwists to an extent that varies with the amount of current applied to it. Therefore, the amount of light passing through the liquid crystal can be controlled by altering the shape of its molecules. LCD technology was initially applied to devices such as watches and electronic calculators with small displays containing a limited number of alphanumeric and mathematical characters, along with small single-cell graphics. The liquid crystals in these basic displays act like shutters, allowing light to pass through them or blocking out the light. Some LCD displays use no illumination of their own but employ a rear mirror to reflect incoming light back through the crystals, while others are backlit to aid viewing.

Dazzling Displays

LCD technology really took off when electronics companies saw their enormous potential use in portable computers. Color LCD monitors are made up of several layers (Figure 2.24). At the rear, a white backlight – usually fluorescent tubing around and/or behind the display – provides the source of illumination. This light passes through a polarizing filter to align the light waves in one direction. A layer of liquid crystals is sandwiched between two glass alignment layers oriented at 90° to one another. Each pixel in the display comprises three sub-pixel cells with red, green, and blue filters. After passing through the filters, the light exits the screen through another polarizer, arranged at right angles to the first polarizing layer. Different colors are produced by illuminating the sub-pixels in various combinations – red and green producing yellow, red and blue producing purple, all three producing white, the absence of all three producing black, and so on (Figure 2.25). When current is applied to each cell, the liquid crystals untwist, allowing various amounts of light to pass through the external polarizer to the viewer; the brightness of each color is controlled in this way (Figure 2.26).

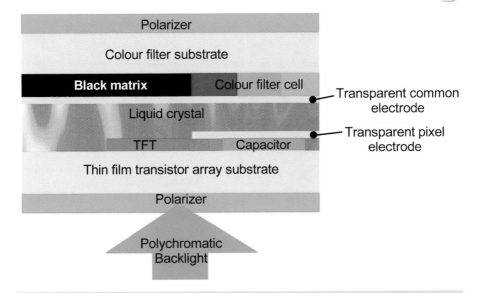

Figure 2.24. Simplified cross-section through a TFT-LCD screen (photo by Peter Grego).

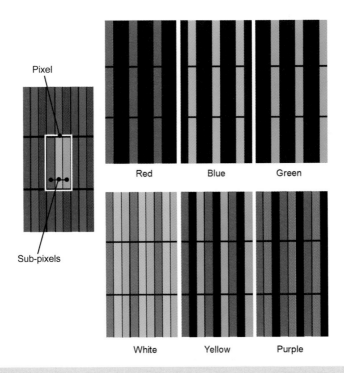

Figure 2.25. Extreme close-up of an LCD screen showing how colors are created from combinations of the *red, green,* and *blue sub-pixels* (photo by Peter Grego).

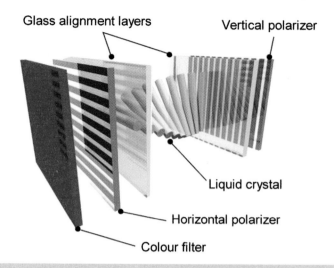

Glass alignment layers · Vertical polarizer · Liquid crystal · Horizontal polarizer · Colour filter

Figure 2.26. The configuration of an LCD sub-pixel (Marvin Raaijmakers, Wikimedia Commons).

Passive Resistance

Many LCD displays are of the 'passive matrix' sort, and they were common in most laptops up to the mid-1990s. Passive matrix LCDs have their active electronics (transistors) outside of the screen, and each sub-pixel in the display is activated by means of a current sent along its own particular column and grounded along its own row. This is the reason why older LCD screens are small and have wide bezels. One major drawback of passive matrix screens is that their brightness appears to fall dramatically beyond a certain viewing angle, usually around 30° on either side. In addition, they suffer from poor contrast and a slow response time. You can see this when the cursor is moved rapidly across the screen, causing it to 'submarine' (when the cursor disappears and reappears at intervals).

LCDs Get Active

Most modern computer monitors and laptops have 'active matrix' displays, also known as TFT-LCD (thin film transistor LCD) monitors. Active matrix screens have brighter, far sharper, and more contrasty images than their older passive matrix counterparts. Since each colored sub-pixel on a TFT-LCD screen is controlled by its own transistor (which takes up only a relatively small area), there are a staggering 2,359,296 transistors in a 1024 × 768 color TFT-LCD screen. There are 256 shades of each sub-pixel's color, and these can be obtained by controlling the voltage applied to them by each transistor, producing a total of no fewer than 16.8 million possible colors. TFT-LCD monitors have a much better range of viewing angles than passive matrix monitors, the best offering up to 160° of viewing comfort.

Problem Pixels

Sometimes, individual LCD pixels don't behave as they ought to. There are three flavors of defective pixel: hot pixels (always switched on), dead pixels (that cannot be turned on), and stuck pixels formed from one or two hot or dead sub-pixels. Although defective pixels are very small things that take up only a minute fraction of the screen area, they can show up as plain as a pikestaff, especially if they are positioned away from the screen's edge. Prospective purchasers of used equipment featuring LCDs (digicams, PDAs, laptops, and computer monitors, for example) would be wise to note whether problem pixels are mentioned in the advertisement. Some sensitive folk may find the constant presence of a tiny rogue dot of light or shade as annoying as a bug on a television screen.

Different Aspects

Monitors with the traditional 4:3 aspect ratio are now being outnumbered (in terms of production) by wide-screen 16:9 aspect ratio TFT-LCD monitors (Figures 2.27 and 2.28). For graphics work you may find that the older aspect is more comfortable to work with, although you can have two TFT-LCD monitors on your desktop, such as a 22-in. wide-screen and a 21-in. 4:3. Interestingly, the latter has the largest screen area by 4 in.^{2}, owing to its 4:3 aspect ratio.

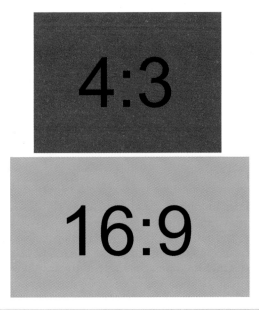

Figure 2.27. Standard 4:3 aspect ratio screen and 16:9 wide-screen compared (photo by Peter Grego).

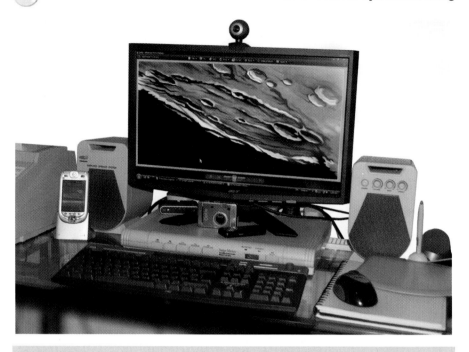

Figure 2.28. This author's 22-in. wide-screen TFT-LCD monitor (photo by Peter Grego).

The popularity of TFT-LCD wide-screens is largely the result of the wide-screen format of most modern movie DVD–video releases and many television programs. It's worth pointing out that there can be image distortion problems when wide-screen monitors are used to display certain programs designed for the 4:3 format (especially older software), or when wide-screen monitors are plugged into some older graphics cards. If problems are encountered when a new wide-screen monitor is used with an older computer, then it's probably time to upgrade the graphics software, the graphics card, or the computer itself.

Printing and Scanning

It goes without saying that printers and scanners are important peripherals for just about any computer user, let alone someone interested in cybersketching and doing graphics work on the computer. Viewing cybersketches on a good monitor is a pleasant experience, but printers create a hard copy, enabling the cybersketcher to enjoy a tactile, portable, permanent physical record of his or her electronic graphic endeavors. Prints can be incorporated into observing logbooks, compiled into albums of their own, shared with others through regular 'snail mail,' and assembled into display material for astronomical exhibitions. Scanners are useful for digitizing printed images and regular observational drawings, allowing them to be stored on a computer and subsequently changed or enhanced to improve them; once digitized, graphics can, of course, be stored, duplicated, and distributed electronically.

Printworks

If you were a computer user during the 1980s, you'll doubtless be able to recall the screeching din caused by your first dot matrix printer (DMP). DMPs were the most common form of computer printer throughout the decade. As their name suggests, their output is comprised of matrices of small printed dots which are produced by the impact of small pins against an ink-coated ribbon. Color DMPs use different colored ribbons (black, red, yellow, and blue), mixing the patterns of colors to produce a range of hues and shades. DMPs produce low-resolution text and images (the dots being visible with the unaided eye), and today they are most commonly used in businesses for receipt printing purposes.

Inkjet printers superseded DMPs in the early 1990s, and they have continued to enjoy tremendous popularity among personal computer users. Nozzles in the inkjet printer's head create precise patterns of tiny ink droplets on paper to a resolution greater than that visible with the unaided eye, with resolutions up to 1440×720 dots per inch (dpi). Mixtures of black, red, yellow, and blue ink produce a huge range of hues and tones. Most inkjet printers have separate cartridges for the black and color inks, and in some printers each color has its own cartridge.

Inkjet printers come in two types – those with print heads fixed inside the machine and those whose print heads are attached to the disposable ink cartridges (Figures 2.29 and 2.30). Since fixed print heads are supposed to be permanent features that last for the life of the printer, the ink cartridges for these printers need only serve as reservoirs, so the cost of consumables is kept down. Because of the relative physical simplicity of these cartridges, it's often a viable option to refill them yourself (or at an ink shop) to further reduce running costs. Fixed head printers usually require no calibration (precisely adjusting the jets of ink so that they are perfectly aligned) each time the ink is changed. One major drawback lies with the fact that if the head becomes faulty in some way then it is often more economical to purchase a new printer than to have the problem fixed (unless the printer is under warranty and the problem can be fixed at little cost).

Inkjet cartridges featuring their own integral print heads are more expensive consumables. Each time the ink runs low it becomes necessary to refill the cartridge or dispose of it, print head and all. Some high-end printers reduce running costs considerably by having disposable print heads that are not permanently attached to the ink cartridges; both cartridge and print head can be separated and replaced when necessary.

Originally developed in the 1970s, laser printers rapidly gained in affordability and popularity from the 1990s onward. Contrary to popular belief, laser printers don't somehow burn an image onto paper by shooting a laser beam at it. Instead, the image to be printed is encoded into rasterized data by the

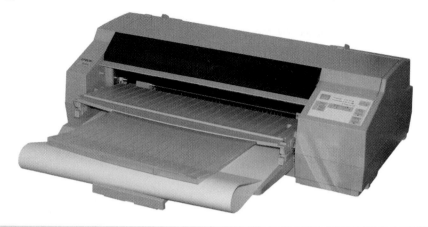

Figure 2.29. The author's large format Epson Stylus Color 1520 inkjet printer has a fixed print head (photo by Peter Grego).

computer (broken up into a grid of pixels), and the eponymous laser represents this data by switching on or off according to the color of the image pixel. The laser beam encoded with image data is guided, by means of a scanning mirror and a guiding lens, across the surface of a photoreceptor – a rotating cylinder or belt that is negatively charged in the dark. Wherever the laser beam hits the photoreceptor, it loses its negative charge, and in this way an electrostatic image is built up on its surface. Laser toner is a fine powder containing color mixed with tiny plastic particles, and it is attracted to static electricity; the

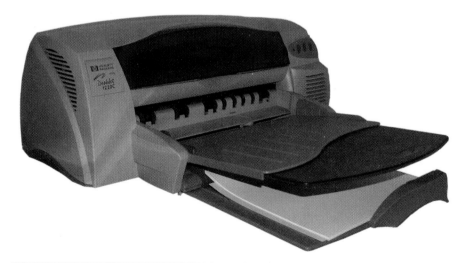

Figure 2.30. The author's A3 format Hewlett-Packard Deskjet 1220C inkjet printer. It uses cartridges with their own print heads (photo by Peter Grego).

toner clings to those parts of the photoreceptor that retains its negative charge, leaving the areas exposed by the laser uncoated. As the photoreceptor is rolled across a sheet of paper, the toner image is transferred to the paper. Finally, the paper is passed through a high-temperature fusing roller (up to 200°C), which melts the plastic component of the toner, permanently fixing the image to the paper. Laser printers offer fast, sharp, smudge-free, high-resolution prints that are particularly useful as field templates for astronomical sketches (see later in cybersketching techniques) (Figure 2.31).

Figure 2.31. The author's Canon CLP-300 color laser printer (photo by Peter Grego).

Scanners

Image scanners allow 2D text and graphics to be digitized and stored electronically. Now a standard peripheral with most computer systems, scanners have come a long way since the first one was developed more than half a century ago at the US National Bureau of Standards by a team lead by Russell Kirsch (Figure 2.32).

Flatbed scanners are by far the most common type of image scanner (Figure 2.33). The graphic to be scanned is placed facedown and in contact with a platen (clear pane of glass), excluding as much external light as possible. A bright fluorescent light source beneath the platen illuminates the graphic, and a moving

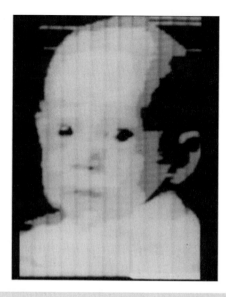

Figure 2.32. This black and white picture of scientist Russell Kirsch's 3-month-old son, Walden, is the first picture ever scanned. It has a resolution of 176 × 176 pixels (Russell Kirsch, Wikimedia Commons).

optical array containing a set of CCD sensors (actually, three arrays, with red, green, and blue filters in color scanners) reads the illuminated area and digitizes the image according to the brightness and color of the light being reflected from the graphic. High-gloss images sometimes present a problem in scanning because of their tendency to reflect the source of illumination straight back into the sensors, but this is more a problem for older scanners.

Handheld image scanners were once widely available during the early 1990s; they operate along the same lines, usually using a line of LEDs as their source of illumination. The user is required to manually sweep the device as straight and as even a speed as possible over the graphic. Needless to say, handheld scanners produce images that are prone to distortion and other artifacts, but they are useful for capturing images from sources that are difficult to scan on a conventional flatbed device. For example, you can use a handheld scanner to capture astronomical images from the pages of antique books whose spines and delicate binding simply couldn't abide being squashed up to the platen of a flatbed scanner.

Scanning technology isn't just restricted to capturing text, drawings, and graphics on paper. Film slides and negatives can also be scanned and digitized, but they require illumination from the top side, rather than from beneath, in order for their details to be seen by the optical array. Some flatbed scanners have their own built-in slide/transparency scanners, and inexpensive adapters for use with regular flatbed scanners are also available. However, the very nature of the subject matter of astronomical images often means that slides and negatives of astrophotographs require a little more from a scanner in terms of

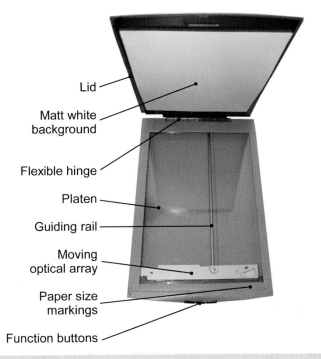

Lid
Matt white background
Flexible hinge
Platen
Guiding rail
Moving optical array
Paper size markings
Function buttons

Figure 2.33. Main components of the author's Canon CanoScan LiDE 30, a USB-powered flatbed scanner (photo by Peter Grego).

sensitivity and resolution than do images of regular, well-illuminated everyday subjects. Special slide and film scanners are capable of scanning 35 mm film and mounted slides, in addition to other film formats, typically to resolutions of 3,600 dpi or greater.

Scanning software usually allows a quick, low-resolution preview of the image to be displayed on-screen so that the user can more precisely select the area to be scanned. Prior to scanning, it's usually possible to set the desired resolution (within the scanner's range), the type of image being scanned, whether it's a halftone print (as in newspaper images), a regular photograph or document, and the color mode (black and white, grayscale, or color). Scanned images can be saved in a number of graphical formats, the most common being BMP, JPEG, and TIFF.

A Few Scanning Tips

- Install the scanner using its own proprietary software.
- Keep the platen dust-free, regularly cleaning it with a moist cloth and drying it with light pressure using a clean cloth.
- Staples and paper clips may scratch the platen's surface. Try to avoid placing them in contact with the platen.

- Use the buttons on your scanner to save time, and assign desired functions to them.
- Align the page with the top edge of the platen and within the marked boundaries of the paper size. Time is wasted if, for example, a small image is placed at the bottom of the platen so that much of the time is spent scanning a blank space.
- Never press down hard on the platen's surface. Apart from the possibility that the glass might break, any distortion in the shape of the scanner may affect the movement of the optical array.
- Scan artwork to a high resolution to preserve detail. About 300 dpi is an optimum resolution for most work. Don't adjust the scan resolution settings to surpass those of your scanner.
- Make sure that you scan at the brightness and highest contrast settings. It is possible to over- or underexpose a scanned image.
- Scanning software other than that supplied with the device may be more user-friendly, more versatile, and produce better results. Test other scanning software by downloading trial programs, and try before you buy.

Digital Imaging – A Useful Cybersketching Tool

A CCD (charge-coupled device) is a small, flat chip – about the diameter of a match head in most digital cameras – made up of an array of tiny light-sensitive pixels. CCD chips in low-end digicams, including those featured in a number of PDAs, may have an array of 640×480 pixels; a more expensive digicam may have a 2240×1680 pixel CCD chip (around 4 megapixels), while a high-end digicam may boast a chip of more than 8 megapixels. When light hits a CCD pixel it is converted into an electrical signal whose intensity corresponds with the brightness of the light that struck it. This information is then processed into an image and stored in the camera's own flash memory or on a removable flash card.

Electronic imaging devices have revolutionized astronomy in both amateur and professional arenas. Thanks to the little CCD chip, amateur astronomers of relatively modest means can capture stunningly detailed images of the Moon and planets. Faint deep sky objects once considered the visual 'property' of big professional observatories are now within easy reach of the amateur astronomer's sensitive electronic eyes.

Digital imaging is often said to be in competition with visual observation. This is a sentiment expressed by those who don't really understand the nature of amateur astronomy and the rewards to be found in applying and mastering age-old observational techniques. Those same people might look down on those unfortunate mortals who enjoy sketching what they see through the telescope eyepiece, confident in the belief that their unblinking chips are the way of the future and that science has left the visual observer and sketcher of things celestial to revel in an arcane pursuit of relevance only to a dwindling few of artistic bent. However, astronomical sketching has as much value and relevance today as it did before the first CCD chip bathed in its first photons.

One potent method of cybersketching, particularly applicable to lunar and solar observations, is to use a live or almost-live digital image taken with a digicam, DSLR, webcam, or astronomical CCD device as the basis for an observational drawing. This kind of sketching technique is a 'simultaneous cybersketch' – an electronic drawing made with the aid of a simultaneously displayed digital photographic image that is used either as a reference and/or as a template. Alternatively, it may be a traditional pencil sketch made using a low-contrast printed digital image template – a 'cybertemplate,' if you will – and these kinds of observational drawing might be termed 'digitally assisted' sketches. The techniques involved in these cybersketching methods are discussed in more detail later.

Digicams

Most low-end digital cameras have a fixed optical system, with non-removable lenses, and some may not even have an LCD display. Astronomical photography through these cameras must be done afocally, by aligning the digicam lens with the telescope's eyepiece (Figure 2.34).

Figure 2.34. The author's Pentax Optio S30 digicam, set up for afocal imaging through a 100-mm refractor (photo by Peter Grego).

As a general rule, the digicam with the highest pixel rating will produce the clearest, highest resolution images. Most digicams offer a number of resolutions.

Figure 2.35. The Optio S30, a 3.2 megapixel compact digicam (photo by Peter Grego).

A low-resolution image will take up less space in the camera's memory, allowing more images to be stored, but their quality will be poorer. Digicams provide instant results, and the images can be viewed on the camera's small LCD screen (if it has one). LCD screens are usually on the small side, and the display is coarser than the captured image itself. Each image can be individually reviewed to decide whether it's good enough to keep or whether to delete it and free the memory for a better image.

Because of their basic nature, most digicam astrophotography is restricted to big, bright subjects full of detail, such as the Sun and the Moon. Digicams are designed for everyday use, and their automatic settings may pose considerable problems when attempting astrophotography, so it's essential to experiment with the digicam's various settings to produce the best results. One of the most important settings to come to grips with is the digicam's exposure settings, as many afocal lunar images tend to be overexposed, the bright part of the Moon appearing washed out and lacking in any detail. The digicam's automatic exposure works best if there is a uniformly bright image across the entire field. A digicam may judge exposures perfectly fine when the Moon is centered in the field of view or when taking close-up shots.

Color images of the Moon taken with digicams may show vivid hues that can't be seen visually through the telescope eyepiece. While color can enhance the aesthetic quality of an image, it can also be undesirable. Computer processing by reducing color saturation levels can easily tone down an image. Capturing the Moon's colors in an exaggerated or visually realistic fashion may produce a pleasing image, but the same amount of topographic detail can be recorded in black and white. If your camera has a facility to take black and white images, try it out; the results may be noticeably sharper than those taken in color. Black and white images will also take up less space in the camera's memory, too.

Using the digicam's zoom facility (if it has one) can eliminate the problems of image vignetting that tends to plague afocal photographs. Zooming adjusts the position of the camera's internal lenses; the magnified image becomes progressively dimmer, and vibrations in the telescope will show up more. When 'digital zoom' comes into play at high magnifications, the quality of the image begins to degrade, and the advantages of zooming are completely canceled out. Note that optimum zoom is not the same as maximum zoom. The best amount of zoom to use depends on the seeing conditions, the resolving power of the CCD chip, and the telescope, as well as the stability of the system and the accuracy of the telescope drive.

A low-end digicam is perfectly capable of producing an astronomical image – a 'cybertemplate,' if you will – for use as the basis for an observational drawing of a small area of the Moon or an interesting region of the Sun. It's not necessary for a lunar or solar cybertemplate to be perfectly framed, gloriously detailed, and as sharp as a pin to be used in a digitally assisted sketch. All that's required are the most basic visual prompts to the main features under observation; the observer can fill in all the detail to whatever level is desired using his or her own drawing skills and visual acuity. One of the main practical points of using a cybertemplate is that the observer need not spend a great deal of time in laying out all the broad detail by placing features in their correct relative size and positioning them accurately with respect to one another. This is already done, and all that the observer need do is delineate these features boldly and concentrate on the finer detail.

Mobile Phone and PDA Cameras

The first digital cameras to be incorporated into mobile phones and PDAs were of low resolution and very limited in their capabilities compared to even the most modest digicams. Digital cameras were also developed as add-ons to PDAs, fitting into their CF or SD slots. Still, even a low-resolution mobile phone or PDA camera is capable of being used to take afocal images of the Moon of good enough quality to be used as cybersketching templates (Figure 2.36). Pretty good digital cameras, capable of taking reasonably good images, are now featured on some of the latest mobile phones and PDAs.

Some sort of adapter is required to align the lens of the phone or PDA with the telescope eyepiece and hold it there firmly. One primitive method is to use a nice sticky piece of blu-tack, which does the job well enough for most purposes. Some mobile phones and PDAs can be fitted with a custom-made telephoto lens (Figure 2.37). If this lens isn't good enough for astronomical use in itself, it can be converted to fit into the telescope's eyepiece adapter, providing a secure means of fixing the phone in position for astrophotography.

HP Pocket Camera
(CF slot)

HP Photosmart
Mobile Camera (SD slot)

SPV M2000
Built-in camera

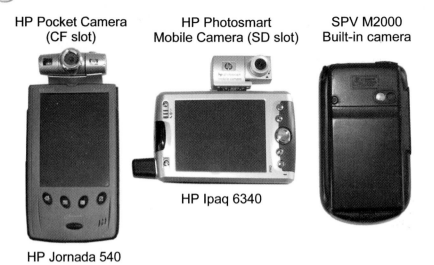

HP Ipaq 6340

HP Jornada 540

Figure 2.36. Three of the author's PDAs, ready for imaging. Each is capable of taking still and/or video footage (photo by Peter Grego).

Figure 2.37. A telephoto lens for mobile phone – which some might consider a useful gadget in itself.

Digital SLRs

Digital single lens reflex cameras (DSLRs) are more capable and versatile than their compact cousins, generally offering higher resolution images, capable of taking time-exposure photographs, and offering a greater range of fine adjustments. The DSLR camera body is attached to the telescope using a T-mount adapter, effectively transforming the telescope into a large telephoto lens (Figure 2.38). Using prime focus, a telescope with a focal length of under 2,000 mm will project the entire half-degree-wide lunar disc onto the camera's CCD chip. Some DSLRs, such as the Olympus E-300 (of which this author has considerable experience), don't offer a live preview on their LCD view screens, so focusing needs to be performed by squinting through the camera's viewfinder. This may be easy enough for the Moon, but not so easy for dimmer objects. Here's where a full-aperture focusing mask comes in handy, using it to merge the image of a bright star to produce optimum focus.

Figure 2.38. The author's Olympus E-300 DSLR, shown with a T-adapter for prime focus imaging through a telescope (photo by Peter Grego).

Camcorders

Since camcorders have fixed lenses, images must be obtained afocally through the telescope eyepiece, and the same challenges that affect afocal imaging using digicams apply to camcorders (Figure 2.39). Camcorders tend to be somewhat heavier than digicams, so it's important that the device is aligned with the telescope eyepiece in as sturdy a fashion as possible. Lightweight digicams and camcorders can be held to the eyepiece using lightweight camera brackets that fit into the eyepiece holder or clamp around it in some fashion. Digital camcorders are the lightest and most versatile of their type; images taken with these can be easily transferred to a computer, either by using a USB link or by transferring the memory card, and the footage can be digitally edited using the same techniques as images obtained with a webcam (see below). Once downloaded onto a computer, single frames from digital video footage can be sampled individually (at low resolution) or stacked using special software to produce detailed, high-resolution images.

Figure 2.39. A camcorder set up for imaging through the telescope eyepiece (photo by Peter Grego).

Camcorder footage often conveys the striking impression of actually observing through the telescope eyepiece. The viewer is using the same cerebral processing as in a real observation, as atmospheric shimmering distorts the view and the eye fixes upon individual objects to attempt to make out the fine detail. Camcorder footage – particularly of the Moon – can be used as a basis for a cybersketch. Live camcorder images can be fed directly to a monitor indoors, allowing a comfortably seated observer to sketch part of the view, be it a lunar crater or a sunspot group. There's no suggestion that this is 'cheating' in some way. Remote though the telescope may be, and though the image is channeled through an electronic sensor, it's real observing nonetheless. Similarly, a pre-recorded video sequence may be played in a continuous loop on an indoor television or monitor and used as material to sketch from. Like observing at the eyepiece, the eye will take advantage of moments of good clarity to gather all the finer details. One obvious advantage of sketching from a video loop is that cloud will never threaten the success of the observing session and there's no chance that cold, numb fingers will ruin the observer's enjoyment and the accuracy of the finished sketch.

Webcams

Although they are designed for use in the home to facilitate communication between individuals over the Internet, webcams can be used to capture high-resolution astronomical images. Costing just a fraction of dedicated astronomical CCD cameras, webcams are extremely lightweight and versatile. Any commercial webcam hooked up to a computer and a telescope can be used to image the Moon and the brighter planets. This can be achieved afocally, by eyepiece projection or by prime focus imaging (Figure 2.40). Electrical signal noise greatly hampers the use of webcams to capture faint deep sky objects such as nebulae and galaxies, but the Moon and major planets are bright enough to produce super images.

Webcams are used to record video clips made up of hundreds, or thousands, of individual images, giving them a distinct advantage over a single-shot astronomical CCD cameras. By taking a video sequence, the effects of poor seeing can be overcome by selecting (either manually or automatically) the clearest images in the clip by means of computer software. These images can then be combined using stacking software to produce a highly detailed image; this may show as much detail as a visual view through the eyepiece using the same instrument.

Most webcams are able to record image sequences using frame rates of between 5 and 60 fps (frames per second). A 10-s video clip made at 5 fps will be composed of 50 individual exposures and may take up around 35 MB of computer memory. At 60 fps there will be 600 exposures, and the amount of space taken up on the hard drive will be proportionately greater. Using the webcam's highest resolution (in most cases, an image size of 640 × 480), a frame rate of between 5 and 10 frames per second and a video clip of 5–10 s is

Figure 2.40. Prime focus imaging at the telescope using a Philips ToUcam PCVC740K webcam (photo by Peter Grego).

optimum. If 5–10 of these clips, centered on the same area, are secured in quick succession, the imager will have between 125 and 1,000 individual images from which to work. The sheer number of images provided by webcams is their greatest strength. A single-shot dedicated astronomical CCD camera costing perhaps ten times as much as a webcam only takes one image at a time. An image produced by an astronomical CCD may have far less signal noise and a higher number of pixels than one taken with a webcam, but in mediocre-seeing conditions, the chances that the image was taken at the precise moment of very good seeing are small. Webcams can be used even in poor seeing conditions, as a number of clearly resolved frames will be available to use in an extended video sequence. Video sequences are usually captured as AVI files.

Astronomical image editing software is used to analyze the video sequence. In addition to commercial software, there are a number of excellent freeware imaging programs available. Some programs are able to work directly from the AVI file, and much of the process can be set up to be automatic. The software itself selects which frames are the sharpest, and these are then automatically aligned, stacked, and sharpened to produce the final image. If more control is required, it is possible to individually select which images out of the sequence ought to be used to compile the final image. This may require up to a 1000 images to be visually examined, so it can be a laborious process, but it can produce sharper images than those derived automatically. Images can be further processed in image manipulation software to remove unwanted arti-facts, to sharpen the image, enhance its tonal range and contrast, and to bring

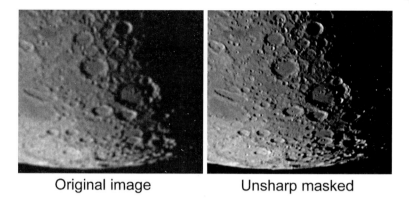

Original image   Unsharp masked

Figure 2.41. The effects of applying unsharp masking to an individual image frame (photo by Peter Grego).

out detail. Unsharp masking is one of the most widely used tools in astronomical imaging; almost magically, a blurred image can be brought into a sharper focus (Figure 2.41). Too much image processing and unsharp masking may produce spurious artifacts in the image's texture and a progressive loss of tonal detail.

CHAPTER THREE

The Power of the Portable

Going by the name of Osborne 1, the first machine to really be considered a portable computer (although you would need pretty big thighs to call it a laptop) was produced by Osborne Computers in 1981. Its 125-mm monochrome CRT screen was sandwiched unceremoniously between two hefty 5.25-in. floppy drives, and the thing was provided with a modem port. Weighing 10 kg and coming with a $1,795 price tag, this sewing machine-sized computer came bundled with a collection of software, and its optional battery gave around 2 h of power. A couple of years later Radio Shack's TRS-80 Model 100 was released, a more manageable and modern-looking laptop weighing 8 kg, but with only a small low-resolution monochrome LCD display. The first true notebook-style laptop with a 4:3 LCD display appeared in 1989 in the guise of the NEC UltraLite (running the MS-DOS operating system), and later that year Apple released a competitor in the form of the first Macintosh portable.

Power in Your Lap

Since those early days, laptop computers (often called notebook computers) have developed into sleek, lightweight machines that in many respects rival the best of their desktop counterparts. Modern laptops come in a variety of types, depending on their size and weight:

- Subnotebook: Less than 225 mm wide by less than 175 mm deep, less than 30 mm thick, and weighing less than 1 kg;

P. Grego, *Astronomical Cybersketching*, Patrick Moore's Practical Astronomy Series, DOI 10.1007/978-0-387-85351-2_3, © Springer Science+Business Media, LLC 2009

- Ultraportable: 225–275 mm wide by 200–250 mm deep, less than 30 mm thick, and weighing 1–2 kg;
- Notebook: 275–350 mm wide by less than 275 mm deep, 25–35 mm thick, and weighing 2–3 kg;
- Large laptop: More than 375 mm wide by more than 275 mm deep, more than 35 mm thick, and weighing more than 3 kg.

A hinge along the edge of laptops at the base of their screen allows them to be opened up (most to 180°, producing a large flat shape), but some can be folded to 360° and/or swiveled according to requirements.

Early models of laptop usually had a trackball pointing device to move the cursor on-screen; these machines consisted of a small recessed ball at the base of the keyboard that could be moved by fingertip (in effect, an inverted mouse), flanked by buttons on either side corresponding to mouse buttons. Some modern laptops, however, use a touchpad to control the movement of the cursor (Figure 3.1). Most touchpads work by means of a matrix of sensors embedded beneath the touchpad's mylar surface; these sensors operate using the principle of capacitance. When two electrically conductive objects approach each other without actually touching – the touchpad's grid and the human finger, separated by the mylar surface – capacitance is created by the interaction between their electrical fields. Measurements of the change in capacitance in the grid allow the computer to detect where the finger is touching; in some touchpads this corresponds to an accuracy of 1 μm. Touchpads can also be programmed to simulate mouse clicks, rather than the user pressing the mouse buttons adjacent to the pad; however, regular finger movements over the pad are sometimes misread as mouse clicks, leading to some annoying temporary disruptions in the flow of work.

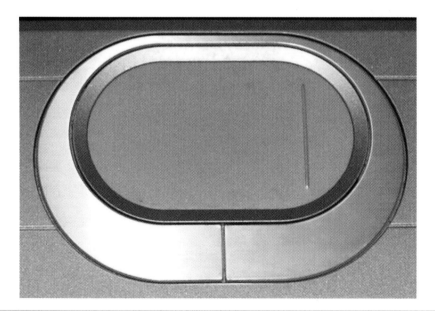

Figure 3.1. Touchpad on the author's Packard Bell EasyNote R4 laptop computer (photo by Peter Grego).

Laptops can be used in the field near the telescope eyepiece and in the observatory for astronomical work, including the use of planetarium-type programs, researching data on celestial objects, imaging and image processing, preparing observing blanks, astronomical sketching using a graphics tablet, or for processing and enhancing digital sketches made on other devices. It's worth bearing in mind, however, that most laptops are not constructed with astronomical field work in mind; under cold, damp, or humid conditions in the field failures of various sorts can occur, although these failures are usually of a temporary nature. Most laptops operate perfectly well between temperatures of around 5–35°C. Subzero temperatures can cause internal condensation to freeze, leading to expansion that may permanently damage the CPU and electronic components. At the other end of the temperature scale, a computer's built-in cooling system may not be able to cope under very warm ambient temperatures, and the CPU (most of which become very warm under normal operational conditions) may seriously overheat, causing the machine to shut itself down. It goes without saying, too, that water and electronics are not the best of friends; a damp (let alone rained-upon) computer may not only suffer permanent damage but the operator may be at serious risk of electrocution.

The author of this book uses a 1998 vintage Compaq Presario 1200 laptop for a number of astronomical tasks in the field, including webcam imaging and operating astronomical programs useful for planning observations (Figure 3.2). This laptop may have an obsolete 12-in. passive matrix display, it may be rather slow at times, and the image may be just a little patchy and washed out, but because of its very lack of value – were it to be ruined by getting rained upon suddenly, or should it crash catastrophically to the ground – the loss would be unfortunate but of little monetary consequence.

Figure 3.2. The author's vintage Compaq Presario 1200 laptop has an old-style passive matrix display. It is shown here set up for webcam imaging (photo by Peter Grego).

Special tough laptops are capable of operating under conditions far more severe than those that would cause a regular laptop to faint or conk out altogether. Produced for the industrial, military, and emergency services markets, these rugged devices are usually an order of magnitude more expensive than regular laptops of a similar performance specification. They are made from a combination of shatterproof, shockproof, and heat-resistant materials; their screens are protected by strong scratchproof glass, the units (including their external sockets) are sealed against infiltration by dirt and water, and their hard drives are shock mounted. Among the most popular of their kind are Panasonic's Toughbook series, the Motorola Rugged Notebook, the Dell Latitude XFR series, and the ruggedized laptops in Toshiba's Tecra range. Standards commonly cited in the United States for the durability of these devices include the US Department of Defense's MIL-STD 810, which covers testing against pressure variations, temperature extremes, rain and humidity, fungus, sand, dust, acceleration, shock, and gunfire.

Touchscreens

Touchscreens permit input directly onto a computer's visual display using the fingertip and/or a stylus. Their development led to great changes across a wide arena of computer-related devices, from check-out displays to mobile computing. The first touchscreen with a transparent surface was made in 1974 by Sam Hurst and Elographics (now called Elo TouchSystems). Since then, various touchscreen technologies have emerged, including touchscreens of the resistive, capacitive, surface acoustic wave, acoustic pulse, infrared, optical imaging, and dispersive signal sort, each of which operates along a different principle and each being more suited to certain applications than others.

Resistive touchscreens are most commonly found on PDAs and portable computers (Figure 3.3). They consist of glass or durable transparent acrylic panels coated with electrically conductive and resistive layers; the layers are prevented from coming into contact with each other by invisible separator points. An electrical current propagates through the screen, and this remains uninterrupted until pressure is applied. At the touch of a finger or stylus, the layers come into contact, producing a change in voltage that the computer recognizes as a touch event at a specific point on the screen, this point being determined by its position along the X-axis of one layer and the Y-axis on the other layer. Resistive technology can be applied to CRT as well as LCD screens. The 5-wire (or more) resistive screens have the advantage of determining the point of pressure contact to a very high resolution, providing touch control accurate enough for fine graphic work with a stylus. Resistive touchscreens are fairly durable, although prone to scratch damage from sharp objects, but they are not adversely affected by dirt, moisture, or light. The resistive layer does, however, reduce the clarity of light from the monitor below it.

Figure 3.3. The author's XDA, a PDA with a resistive touchscreen. This image shows a simulation of the slight distortion of the upper layer of the screen (power switched off) when pressure is applied with a stylus. If the unit were switched on, the LCD image beneath would appear undistorted, however (photo by Peter Grego).

Surface capacitive touchscreens are responsive to very light touch; there's no requirement to apply pressure when using them. They consist of a glass panel coated with a transparent conductive material (commonly indium tin oxide), around the edges of which is arrayed a pattern of electrodes that propagates a low voltage field of 'normal' capacitance over the screen. When this field is intruded upon by the close presence of another capacitance field – an active (powered) stylus or a human fingertip touching the screen – a tiny amount of current is drawn toward the contact point, producing a fall in voltage. Since the current flowing from each side of the screen is proportional to the distance to the contact point, its precise location may be determined along the screen's X and Y axes. Capacitive touchscreens offer considerably better clarity than resistive ones, and they are featured on the latest high-end PDAs, such as the Samsung i900 and the Apple iPhone (Figure 3.4).

Most touchscreens offer 256 levels of pressure sensitivity, half the sensitivity provided by most new graphics tablets. This is perfectly adequate for

 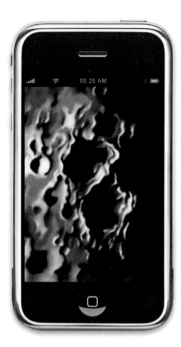

Samsung i900   Apple iPhone

Figure 3.4. Samsung i900 PDA compared with Apple's iPhone, showing the same image on-screen. Note that the iPhone has a larger display area (photo by Peter Grego).

cybersketching. Some new touchscreens, notably those featured on high-end tablet computers, feature 512 levels of sensitivity.

Light reflected from the LCD screen can be annoying when viewed in bright conditions, often rendering the screen unreadable. Some LCD screens attempt to combat this by having a matt surface to scatter reflected light, but these tend to reduce image contrast and color intensity, adding a certain degree of blur and reducing the angle at which the screen may be viewed. Glossy (reflective) LCDs employ a special coating to cut down on the amount of reflected light, and they are particularly effective when used in low-light conditions, such as astronomical observing at night. Glossy screens reflect more light than matt screens, and they tend to deliver a less accurate rendition of grayscale and color tones. If glossy screen reflections are annoying, a screen protector (a transparent plastic overlay that adheres to the screen) may prove to be of some help.

Doing It with Stylus

Slender sticks with a tapering tip, styluses are more akin to the ancient clay tablet cuneiform writing apparatus of ancient Sumeria than the often used alternative word 'pen' (when applied to tablet PCs). Most portable touchscreen devices are provided with their own stylus, which slides into the side, top, or bottom of the unit and is held in place by friction. A number of early PDAs (including the HP Jornada 540) didn't have a stylus slot in the main body of the PDA but instead were inserted into the device's protective flip cover.

It's difficult to credit the fact that styli vary so widely in size, shape, weight, and materials, considering the pretty standard physiological configuration of the human hand and the fact that regular pens have been around for centuries (Figure 3.5). It appears that manufacturers have been so eager to experiment with new stylus designs and/or cut costs that certain sacrifices in usability and user comfort have been made here and there along the way. People's tastes do vary, though, and what one person might consider to be a feeble, almost useless stylus another might think was eminently suited to the purpose. Some PDAs, such as the HP Jornada 540 and HP iPAQ 6300, include in the package the cheapest and most lightweight of plastic styluses as standard; other devices have better, more substantial feeling quality styluses made out of metal and plastic. Some enterprising manufacturers have produced their own device-specific compatible styluses, enabling the user to upgrade their stylus. One model even has an extremely useful built-in ballpoint pen.

Figure 3.5. An array of styluses used in the author's collection of PDAs, P/PCs, H/PCs, and tablet PCs. Key: 1 – Sony Clié PEG-SJ30 (PDA); 2 – Palm Tungsten T3 (PDA); 3 – HP Jornada 540 (P/PC); 4 – O2 XDA (P/PC); 5 – HP iPAQ 6300 (P/PC); 6 – SPV M2000 (P/PC); 7 – SPV M5000 (P/PC); 8 – XDA Orbit (P/PC); 9 – HP Jornada 720 (H/PC); 10 – Fujitsu Stylistic 3400 (tablet PC); 11 – Walkabout HH3 (tablet PC); 12 – ViewSonic V1100 (tablet PC) (photos by Peter Grego).

The very thin nature of the PDA stylus requires a bit of getting used to, especially when inputting handwriting or graphics, but, after a while, it can become almost as comfortable to use as a regular pencil. Having said this, note that smaller devices have smaller styluses, and this makes cybersketching somewhat more difficult. Apart from the fact that drawing on a PDA with a small screen (such as the 2.8 in. screen XDA Orbit) is less satisfying than using a larger screen PDA, the stylus may not be long enough to nestle in the web space between the thumb and forefinger. Not only might the end of the stylus produce a physical source of irritation to the user's palm, the cybersketch is bound to be less accurate. To illustrate this, imagine a pool player who doesn't form a cradle for the cue stick with the back of the fingers, in order to line up the shot, choosing instead to prod the ball with a loose cue.

Styluses for tablet PCs are much more pen-like in size and shape. They are usually made to a good standard (they don't often feel like cheap sticks of plastic), and they are comfortable to hold. The addition of rocker buttons on active styluses (used to click on items) is useful, but these features can sometimes be difficult to use because the pen needs to be rotated and held in the correct manner. The buttons can also get in the way of making a cybersketch. You might sometimes find windows inadvertently opening up on the screen of your Walkabout HH3 tablet computer as the result of accidentally applying pressure to the stylus button.

Tablet PCs

Tablet PCs are extremely versatile portable computers with touchscreen input. Measuring roughly the same size as a sheet of A4, or letter-sized paper, and with the thickness of a matchbox, tablet computers come in three basic varieties: slates, convertibles, and hybrids.

Slates (so-called because they bear a slight resemblance to the writing slates of old) are thin, touchscreen computers without a physically integrated keyboard (Figure 3.6). They usually run full-featured versions of Windows Vista OS or Windows XP Tablet Edition. Prior to the latter, Microsoft released a number of add-ons for Windows called Windows for Pen Computing.

Some slates can be mounted in their own tilt-adjustable desktop cradles so that they can be used as conventional computer monitors; power is channeled through the cradle, and keyboard, mouse, and other peripherals plug into various ports on the cradle.

Ultra-mobile PCs (UMPCs) are a type of slate computer with a highly portable paperback-sized form factor. UMPCs are distinct from PDAs because of their full-featured OS and their consequent ability to run the same programs as a laptop or desktop computer. Slates and UMPCs usually have one or more USB ports, so a mouse and conventional keyboard can be added as alternative input devices. Way beyond most budgets, the latest devices use capacitive technology with palm recognition so that stylus input isn't disrupted by brushing the hand against the screen while writing.

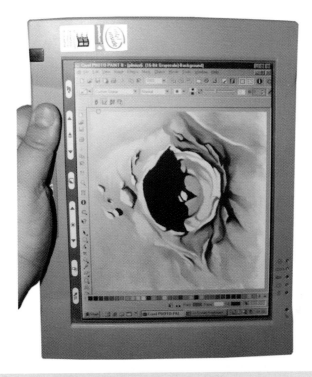

Figure 3.6. A lunar observational drawing made with Corel PhotoPaint, shown on the author's Fujitsu Stylistic 3400, a tablet PC running a pen-enabled version of Windows ME (photo by Peter Grego).

Multi-touch allows two separate input points to be recognized simultaneously, and superbly bright screens can be read in sunlight.

Thin-client slates are network-dependent devices without a large hard drive or optical disk drive or any major programs. They use a network connection (bluetooth or WiFi) to access applications and data on a remote computer.

Thicker and heavier than slates, convertibles are laptop computers that feature a rotating hinge between the keyboard and a touchscreen monitor. This design allows the device to be used as a regular laptop with keyboard input or as a slate when the monitor is swiveled 180° and folded over the keyboard. Their dual nature has made them the most popular form of tablet computer, but the universal hinge has proven to be its major weakness. Some convertibles have dispensed with the hinge and use a sliding screen design.

In an attempt to merge the benefits of slate and convertible computer, hybrids use a detachable keyboard that joins to the monitor with a swivel hinge. Hybrids can be used as a slate (keyboard detached), as a convertible (with the monitor folded over the keyboard), or as a regular laptop. Again, the hinge is the weakest feature – more so because of its need to be used as a point of frequent attachment/detachment.

Some Tablet Comparisons

All new tablet computers can be used for cybersketching. Disappointingly, Microsoft has never performed a proper makeover of Windows Paint, the extremely basic graphics program it has bundled with the Windows OS for decades. Even in Vista, Windows Paint remains the same bare-bones drawing program (perhaps it ought to be renamed 'Windows pre-school daub'). Thankfully, infinitely more capable graphics programs are available as freeware downloads or commercially (see later for details about graphics software). You will need one of these programs to develop your cybersketching skills.

The specifications of a number of popular devices are given below. The older devices that the author has used extensively in cybersketching pursuits are included – Fujitsu Stylistic 3400, Walkabout Hammerhead HH3, and ViewSonic Tablet PC V1100. These have performed very well indeed in some pretty adverse conditions.

Fujitsu Stylistic 3400 (Figure 3.7)

> OS: Windows ME
> Processor: Intel Pentium III 400 MHz CPU
> Memory: 64 MB RAM
> Display: 10.4 in. (26.4 cm) 800 × 600, matt screen
> Digitizer: Resistive panel
> Storage: 6 GB hard drive
> Size: 28.4 × 21.6 × 2.9 cm (11.2 × 8.5 × 1.1 in.)

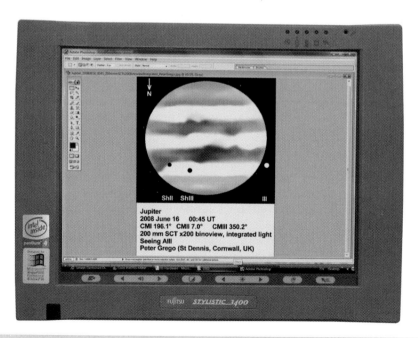

Figure 3.7. The author's Fujitsu Stylistic 3400 (simulated screen image) (photo by Peter Grego).

Weight: 1.4 kg (3.1 lb)

Mobile power: 2.6 AH Li-Ion battery (2.5 h)

Interfaces and connectivity: USB, J11, RJ45 Ethernet port, IRDA, PCMCIA, audio in/out

Notes: A good performer at the telescope eyepiece, lightweight, and comfortable to hold. Its weakest point is the relatively short duration of a charged battery.

Walkabout Hammerhead HH3 (Figure 3.8)

OS: Windows 2000 Professional

Processor: Intel Pentium III 400 MHz CPU

Memory: 130 MB RAM

Display: 10.4 in. (26.4 cm) 800 × 600, matt screen

Digitizer: Capacitive panel (active stylus)

Storage: 10 GB hard dive

Size: 28.4 × 20.1 × 4.1 cm (11.2 × 7.9 × 1.6 in.)

Weight: 1.7 kg (3.7 lb)

Mobile power: 1.8 AH Li-Ion battery ×2 (5 h)

Interfaces and connectivity: USB, IRDA, Type III PCMCIA slot (Type II PCMCIA slots ×2)

Notes: One of the first tough tablet computers, the HH3 features a strong metal alloy body, rubber-sealed units, a shock-mounted hard drive, and a scratch-resistant glass screen. It performs very well in the field for all astronomical applications. With its capacitive screen, the active electromagnetic stylus can move the cursor without touching the screen. A light press of the stylus tip against the screen activates the click; it features an additional stylus button that can be used in right click equivalent mode. Because the special stylus

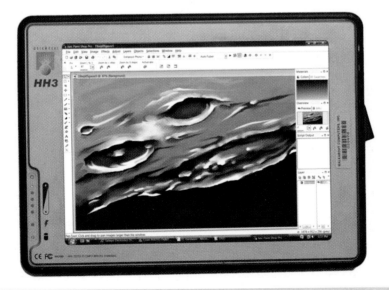

Figure 3.8. The author's Walkabout Hammerhead HH3 tough tablet PC (simulated screen image) (photo by Peter Grego).

alone is used for on-screen input, the hand can be pressed against the screen without consequences. A tough device, the HH3 is capable of sustaining knocks, moisture, and dirt – the promotional video even has a truck drive over it without appearing to damage it. It has an optional LCD and hard drive heaters for cold conditions that operate when the unit is plugged into AC main power. The heaters can be programmed to activate each time the temperature falls below 0°C (32°F). Dual battery means that it can be operated continually by rotating batteries and 'hot swapping' them. Because of the HH3's thick toughened glass matt screen, its image clarity is its weakest point – graphics take on a slightly frosted or grainy appearance.

ViewSonic Tablet PC V1100 (Figure 3.9)

OS: Windows XP Tablet PC Edition
Processor: Intel Pentium III 866 MHz CPU
Memory: 256 MB SDRAM
Display: 10.4 in. (26.4 cm) 1024 × 768, gloss screen

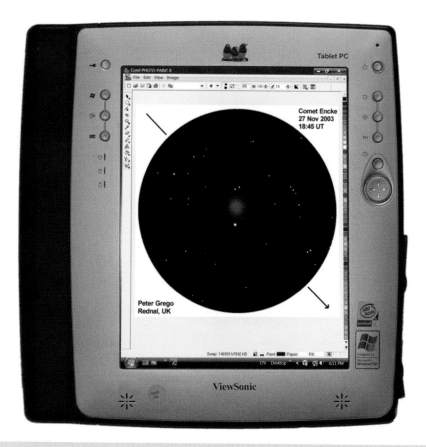

Figure 3.9. The author's ViewSonic Tablet PC V1100 (simulated screen image) (photo by Peter Grego).

Digitizer: Resistive panel

Storage: 20 GB hard drive

Size: 28.8 × 25.3 × 2.9 cm (11.3 × 10 × 1.1 in.) with side battery

Weight: 1.5 kg (3.4 lb)

Mobile power: 4-cell 3.9 AH Li-Ion battery (4 h)

Interfaces and connectivity: Built-in WiFi, RJ45 Ethernet port, Type II PCMCIA slot, Type II CF slot, USB ×2, Firewire (4-pin non-powered), VGA, audio in/out, a 56 k modem

Notes: The V1100 is a good performer in the field for cybersketching. Its built-in WiFi connectivity is a great bonus, enabling Internet access for planning and researching observations and live image webcasting, among a host of other useful things (wireless connectivity is also possible on the Stylistic 3400 and HH3 but only with the insertion of a WiFi card). The V1100 is reasonably light in weight, but the large side battery does not help. If held in portrait mode on the battery side (the way that the device's layout suggests to hold it if you're right-handed), it feels as if the battery might suddenly disengage or snap off.

Newer tablet computers:

Fujitsu Stylistic ST5111 (a slate-type tablet PC) (Figure 3.10)

OS: Windows Vista

Processor: Intel Core 2 Duo 1.2 GHz CPU

Memory: 1 GB DDR II SDRAM

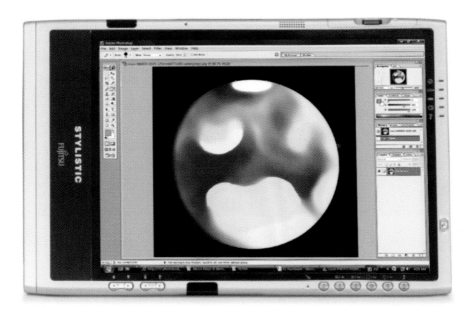

Figure 3.10. The Fujitsu Stylistic ST5111 (simulated screen image).

Display: 10.4 in. (26.4 cm) 1024 × 768, glossy screen
Digitizer: Resistive panel (stylus or finger touch)
Storage: 80 GB hard drive
Size: 32.3 × 22.1 × 2.5 cm (12.7 × 8.7 × 1 in.)
Weight: 1.6 kg (3.5 lb)
Mobile power: 6 cell 5.2 AH Li-Ion battery (6 h)
Interfaces and connectivity: Built-in WiFi, multi card reader, 2× high-speed
     USB, IrDA, RJ11, RJ45 Ethernet, FireWire, audio in/out

Samsung Q1 (an ultra-mobile PC) (Figure 3.11)

OS: Windows XP Tablet
Processor: Intel Stealey A110 800 MHz CPU
Memory: 1024 MB DDR2 RAM
Display: 7 in. (175 mm) WSVGA 1024 × 600, glossy screen
Digitizer: Resistive panel (stylus or finger touch)
Storage: 60 GB hard drive
Size: 22.8 × 12.4 × 2.4 cm (9 × 4.9 × 0.9 in.)
Weight: 0.7 kg
Mobile power: 4 cell Li-Ion battery
Interfaces and connectivity: 2× high-speed USB, multi card reader, RJ45
     Ethernet, integrated Bluetooth and WiFi, novel QWERTY keyboard, digital
     camera

Figure 3.11. The Samsung Q1 UMPC.

CHAPTER FOUR

Handheld Cyberware

During the 1980s a new type of computer called the personal information manager (PIM) appeared, a handheld device whose functions were intended to match the versatile but bulky ring-bound personal organizer of the sort so beloved by 'yuppies' (Figure 4.1). Addresses, telephone numbers, and short text messages could be stored on the PIM's modest flash memory, which also offered a calendar, clock, and alarm function. Data input was through the PIM's small integrated keyboard; some more advanced PIMs allowed data to be input via a PC-link cable connection. Notable among the PIMs was a series of handheld computers developed by the UK company Psion. PIMs are still available, and the line between PIMs and PDAs has been blurred by virtue of the similar form and large color LCD screens (some of them touchscreens) of some devices.

Apple's Newton MessagePad of 1994 is widely considered to be the first true PDA (Figure 4.2). Running the Newton OS, the device featured a monochrome touchscreen upon which freehand 'sketches,' 'shapes,' and 'ink text' could be drawn with the stylus; the shapes could be transformed into vector graphics and even word recognition (though of a notoriously poor 85% recognition rate) was possible with handwritten text. The screen could be rotated through 90° and worked on in landscape or portrait mode – a facility that wasn't commonly offered by many supposedly more sophisticated PDAs for several years after the Newton appeared.

All modern PDAs are small enough to be held in the palm of the hand; they are portable because they have a self-contained power source that can be recharged. PDAs have a graphic display (usually color) with stylus touchscreen input, onboard RAM to store the OS, and other programs, plus removable storage

P. Grego, *Astronomical Cybersketching*, Patrick Moore's Practical Astronomy Series, DOI 10.1007/978-0-387-85351-2_4, © Springer Science+Business Media, LLC 2009

Figure 4.1. PIMs such as this Casio Business Navigator BN-40A may be useful touchscreen devices, but they can't be used for detailed cybersketching (photo courtesy of Wikimedia Commons).

(SD/CF/memory stick). PDAs are linkable to PC or Mac for synchronization, downloading programs, and transferring files and other data.

Like all touchscreen devices, PDAs require calibration with the stylus to ensure a good level of pointing accuracy. This is achieved using calibration software utility provided on the device and is usually completed with several well-aimed taps of the stylus tip on a series of on-screen targets located in several parts of the screen. More care taken during calibration ensures more accurate cybersketching strokes.

Palms

Newton's nemesis came along in the form of the PalmPilot, a PDA launched in 1996 by Palm Computing (then part of the company U. S. Robotics) (Figure 4.3). PalmPilot's LCD touchscreen was monochrome and measured 320×248 pixels; at its base there featured a novel Graffiti handwriting recognition scratchpad, a passive silk screen touchpad on which the user entered simplified (mainly single-stroke) character forms with the stylus to spell out words and enter numerals. Two years later, production of the Apple Newton ended, but it's wrong to assume that these early competitors were technologically superior to the Newton; they simply proved to be more popular because they were smaller and less expensive.

Many Palm OS PDAs that maintained the silk screen Graffiti scratchpad had smaller, squarer LCD touchscreens than their WinCE pocket PC (P/PC) rivals – a

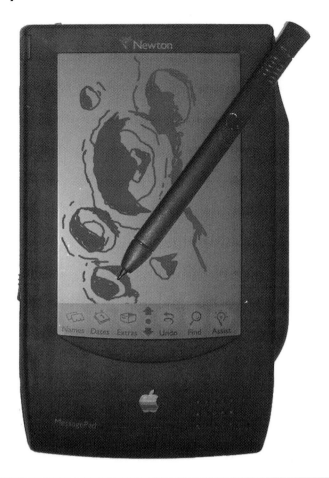

Figure 4.2. Shown here with a simulated astronomical cybersketch, the Apple Newton was the first true PDA (Staeker, Wikimedia Commons).

diagonal screen measurement of around 2.6–3.2 in. (6.6–8.1 cm), compared with around 3.7–3.9 in. (9.4–9.9 cm). Their smaller screen size – around 25% smaller – makes cybersketching (and other tasks) a little more trying on the eyes.

PDAs with the Palm OS introduced a number of innovations, including an electroluminescent backlight for operating in low-light conditions and an infrared file transfer capability (featured on the Palm III in 1998), a color LCD touchscreen (Palm IIIc, 1999), a reasonably high-resolution PDA camera (Sony Clié NZ90, 2002), and an extending LCD screen (Palm Tungsten T3, 2003), the base of which can be used as a 'virtual' scratchpad for Graffiti handwriting recognition (Figure 4.4).

Palm-powered devices have gone through a series of OS modifications over the years, as technology has allowed their capabilities to grow. Newer Palm OS devices are generally more backward-compatible than Windows Mobile OS devices, in that older Palm software will often work without a hitch on newer devices. A brand new Palm OS, based on the Linux OS and code-named Nova, was due for launch in 2009.

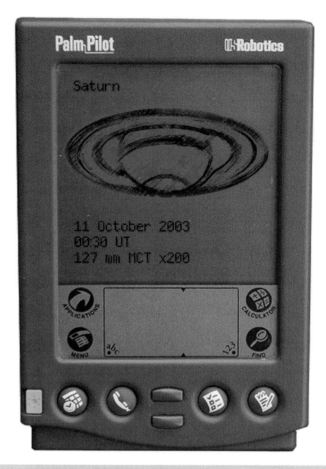

Figure 4.3. An early PalmPilot PDA, showing a simulated astronomical cybersketch (Channel R, Wikimedia Commons).

WinCE

Aware of the success of Palm-powered devices, Microsoft was keen to establish a foothold in the growing handheld computing market, and in 1996 they released the first version of Windows for devices with minimal storage – Windows CE (Windows Embedded Compact post version 6.0). Devices using the WinCE operating system appeared later that year, with Compaq, Hewlett-Packard, Hitachi, NEC, and Philips being the major manufacturers.

Following the launch of WinCE, two basic forms of device making use of the new OS emerged – the Pocket PC (P/PC) and the Handheld PC (H/PC). Pocket PCs are PDAs that fit on the palm of the hand. P/PCs have a 4:3 ratio touchscreen (usually defaulted to portrait mode), a number of function buttons above and/or below the

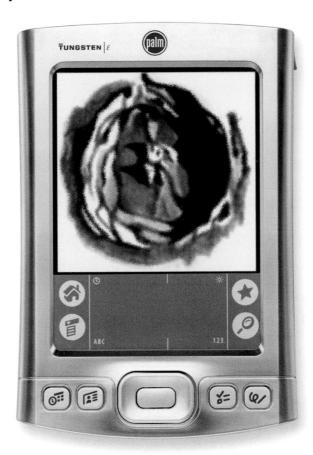

Figure 4.4. A Palm Tungsten E PDA, showing the square LCD touchscreen and the Graffiti scratchpad beneath it (photo by Peter Grego).

screen and on the sides of the device, a slot for a memory card, and sockets for power and computer connectivity (many of them slot into their own desktop cradle to make the latter tasks easier) (Figure 4.5).

Handheld PCs are altogether different beasts. Resembling miniature laptop computers, they incorporate a physical keyboard – usually laid out in the traditional QWERTY fashion – and a long landscape touchscreen. Somewhat confusingly, H/PCs are often referred to as palmtop PCs, to differentiate them from laptop PCs. The Hewlett-Packard Jornada 720 H/PC has a screen with an 11:4 ratio (640 × 240 pixel resolution), making it considerably longer in proportion to its height than a regular 16:9 wide-screen ratio (Figure 4.6). This letterbox-style format makes it a very comfortable device upon which to input text, view spreadsheets, and perform a variety of other office-related tasks. However, at first appearances the long landscape screen doesn't seem to present the ideal sort

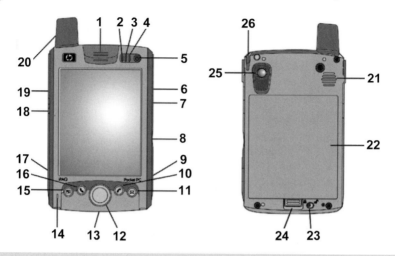

Figure 4.5. The main features of a pocket PC (illustrated by a Hewlett-Packard iPAQ 6300). Key: 1 – Phone receiver; 2 – GSM/GPRS LED indicator; 3 – Bluetooth LED indicator; 4 – WLAN LED indicator; 5 – Power button with charging/notification LED; 6 – Volume up button; 7 – Volume down button; 8 – SD Card slot; 9 – Camera button; 10 – Phone call end button; 11 – Inbox button; 12 – 5-way navigation button; 13 – Charging/communications port; 14 – Microphone; 15 – Contacts button; 16 – Phone send button; 17 – Soft reset; 18 – Voice record button; 19 – Audio connector; 20 – Antenna; 21 – Hands-free speaker; 22 – Removable battery; 23 – Battery lock; 24 – Battery latch (lockable); 25 – Camera lens; 26 – Stylus (photo by Peter Grego).

of space upon which to make cybersketches. However, cybersketching is still possible with H/PCs. In fact, when running a drawings program (usually *Pocket Artist 3*) on the Jornada 720, you can put the cybersketch on one half of the screen, while the other side of the screen (still part of the same image) provides a very useful space for jotting down any notes relevant to the observation, such as the date, time, instrument, seeing conditions, and other data specific to that observation, something that can't really be done on an image in the 4:3 format of a P/PC. Additionally, new images created in *Pocket Artist 3* and a number of other programs will, by default, be matched to the device's screen size (be it a P/PC or H/PC), so the user needn't worry about messing around with the image settings prior to making a cybersketch.

Many PDAs (Palm and P/PC) feature the rather neat ability to rotate the screen contents so that their 4:3 ratio display can be changed from the usual portrait mode to a landscape mode; some even have four display orientations. This function comes in handy for viewing and editing documents, including PDF files, spreadsheets, tables, and the like. However, owing to the layered physical construction of resistive PDA touchscreens, they are optimized for viewing in portrait mode within a certain range of angles, and a distracting unevenness of illumination is visible in several devices when the landscape mode was used. This is supposedly not the case with the newer capacitive touchscreen devices. As far as cybersketching is concerned, it's a matter of personal preference whether the user chooses to depict sketches in portrait or landscape mode; in the former, the image

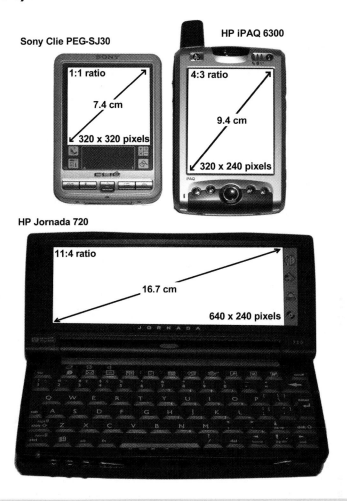

Figure 4.6. Screen formats compared in scale – Sony Clié PEG-SJ30, Hewlett-Packard iPAQ 6300, and Hewlett-Packard Jornada 720 (photo by Peter Grego).

probably looks better on resistive touchscreens, but the PDA is much easier to cradle in the hand when used in this manner.

Like its older sibling (Windows for PC), Windows CE has gone through a number of changes over the years, driven in no small way by advances in technology and the growing capabilities and specifications of P/PCs and H/PCs. WinCE developments include Pocket PC 2000, Handheld PC 2000, Pocket PC 2002, Windows Mobile 2003, Windows Mobile 2003 SE, Windows Mobile 5.0, and Windows Mobile 6.0.

Unlike the OS on desktop computers, the OS on a mobile device can't usually be upgraded to the next OS to come along, but there are downloadable updates available that improve the performance of each particular OS. Programs that work well in one mobile OS may not work at all on a device with a different OS.

Programs are sometimes not even forward compatible. In contrast, programs written for various Windows OS for PC are often forward compatible; for example, some versions of Corel PhotoPaint works just as well in Windows 98 as they do in Windows Vista, even though the program itself actually predates the release of Windows XP.

In the early days of Pocket PC 2000 devices, there were several competing CPU architectures – just as PCs may have Intel or AMD processors – but the difference with mobile devices was that programs sometimes had to be tailored to suit each different CPU. In the first-generation Pocket PC 2000 devices, the Hewlett-Packard Jornada PDA series was equipped with an SH3 processor, the Casio Cassiopeia PDAs had the MIPS processor, and Compaq iPAQs were provided with ARM processors. Since ARM proved to be the best performing of these CPUs, Microsoft decided to standardize the ARM processor for all of the next-generation P/PC devices operating Pocket PC 2002. Later varieties of ARM processor include the XScale, StrongArm, and OMAP CPUs.

PDAs have continued to acquire increasingly more capacious onboard memory and faster CPU speeds. An early PDA, the Hewlett-Packard Jornada 540, operates Pocket PC 2000, uses a 133 MHz Hitachi SH3 CPU, and has 16 MB RAM. A newer PDA, an SPV M2000, operates Windows Mobile 2003 Second Edition, uses a 400 MHz Intel PXA263 processor (an ARM XScale descendant), has 125.77 MB RAM, and 32 MB ROM. My XDA Orbit runs Windows Mobile 6.0, uses a 400 MHz Qualcomm MSM7200 CPU, has 128 MB RAM, and 256 MB ROM. Despite the differences between these PDAs, the Jornada 540 can be used to make astronomical cybersketches – its battery lasts longer, it has a nice, large matt screen for field work, and one of the better cybersketching programs – *Mobile Atelier* – works just as well on this device as it does on the newer ones.

So, when downloading software for the PDA, it's important to know what OS and CPU your PDA uses. PDA program web sites often present a range of download options based on the type of device a particular program caters for. To find out the relevant information of your PDA (along with its RAM capacity), press Start – Settings – System – About.

Most of today's PDAs have high-resolution color glossy touchscreens capable of displaying bright, clear images (visible even in sunshine); many have built-in cameras and may be equipped with telephone, WiFi, Bluetooth, and even GPS connectivity. A number of them (including the SPV M2000) have cool slide-out keypads that illuminate in the dark, making text input a little easier than using the virtual typewriter on-screen (Figure 4.7).

Consumers are demanding increasingly smaller devices, and there is now a merging of the mobile phone with the PDA and camera. Unfortunately, this means that the PDA's screen size is becoming increasingly smaller, making the newer models less suitable for cybersketching. The XDA Orbit PDA is one of these newer, 'cooler' devices; it has a fairly good 3.1 megapixel camera, Bluetooth, WiFi, and quad band phone connectivity, a built-in FM radio, and its own GPS antenna. Its screen, though, measures just 4.4 × 5.8 cm (2.8 in. diagonally) (Figure 4.8).

Faster CPUs in PDAs are desirable up to a point, especially in PDAs where computationally demanding stuff such as GPS might be going on, but increasing processor speed also produces devices that consume battery power at a faster rate. Never leave your WiFi or GPS running while in the field; the batteries drain at an

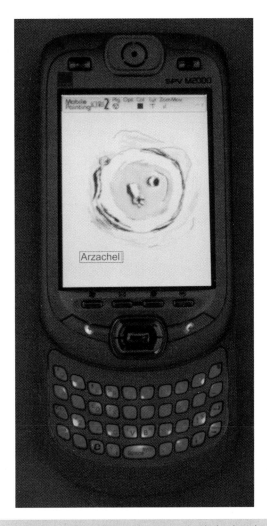

Figure 4.7. The SPV M2000 slide-out keyboard visible in the dark (photo by Peter Grego).

incredibly fast rate, threatening to leave you high and dry in the middle of a cybersketching session.

To summarize, here are a few advantages of using PDAs to make astronomical cybersketches:

- Small and lightweight, easily held in the hand for extended periods of time.
- Long-lasting batteries (typically giving around 3 h of continual use in the field). There are additional portable power options, too, such as bigger capacity batteries and add-on battery packs.
- Backlit, no requirement to use a torch to illuminate a sketch.
- The correct time and date is displayed on the PDA screen – useful information for all astronomical work.

Figure 4.8. The SPV M2000 (3.5-in. diagonal screen) compared with the XDA Orbit (2.8-in. diagonal screen). Cybersketching is easier and more satisfying on the larger screen (photo by Peter Grego).

- Astronomical programs can be used, enabling observations to be researched in the field.
- A wide range of astronomy and graphics programs and utilities are available as freeware, shareware, or commercial software for the PDA.
- PDAs can be used to directly control go-to telescopes.
- Stylus input is intuitive and almost as easy to use as a pencil.
- Many PDAs have an audio recording function, enabling a verbal commentary to be made at the eyepiece to augment a cybersketch.
- Built-in PDA camera allows photography of bright objects afocally through the telescope eyepiece. Images of the Moon can be used as cybersketch templates.
- May accept the same storage card as your digicam, allowing easy transfer of images to be used as cybertemplates.
- Digital templates for observations. The oblate shape of Jupiter or the complex rings of Saturn, for example, can be prepared on PC and transferred to the PDA for use in the field in a cybersketch.
- A Bluetooth- or WiFi-enabled PDA can be used to transfer images directly to or from a PC.

- A PDA with a WiFi connection to the Internet can be used to upload cybersketches to a web site in real time.
- Cybersketches can be easily modified and edited. Because they are digital sketches, multiple copies can be made with no loss of quality.
- No more sharpening pencils, erasing, juggling with torch in hand, etc.

And a few PDA disadvantages:

- Only special 'tough' PDAs are fully waterproof.
- All PDAs are prone to extremes of temperature. They shut down when they get too warm and can suffer irreparable physical damage through the freezing of internal moisture when they become too cold.
- The stylus provided is often very small, awkward to use, and rather easily lost.
- Touchscreen input requires regular calibration to ensure consistent pointing accuracy.
- There's no actual physical copy of the observational drawing; this presents a psychological downside that some people may balk at.

The Outer Limits

Once you have a PDA, and you've been thoroughly impressed with its capabilities, it might be only a matter of time before you consider upgrading to a 'better' device. So, how far do you go? Do you continue to seek out the latest equipment, dumping the older stuff, passing it on to friends, or selling it on eBay when it becomes an embarrassment? Or do you stick with tried and trusted vintage equipment, even though it may perform more slowly and doesn't look as cool as its new counterparts?

Well, it all depends on your budget and requirements, moderated by your personal sense of style and taste and tempered by your attraction to gadgets. If something does its job, doesn't glitch too frequently, and gives you good service in the field, you can't go wrong by sticking with what you have, avoiding the lure of newer technology that may only give you a marginal improvement over your existing equipment. Speaking personally, I don't consider myself a collector or gadget freak by any means, but over the course of the years I've had no fewer than five desktop computers, seven digital cameras, five webcams, three laptops, three tablet PCs, two palmtops and eight PDAs – many of these devices have been discussed in this chapter. In my defense (a defense practiced on many occasions before my wife) many of these devices were bought as used items, and I'm lucky that they have all been fully functioning and have worked as advertised. It's worth noting the point mentioned earlier – my very first PDA, a Hewlett-Packard Jornada 540, gives the newer PDAs a run for their money when using a simple but very capable drawing program such as *Mobile Atelier*, which has a good many years of use left in it.

CHAPTER FIVE

Portable Data Storage

Nowhere is the charming old phrase 'Don't put all your eggs into one basket' more applicable than in the world of data storage. Everyone interested in storing and manipulating images, making digital drawings, taking astronomical images with digicams and CCD cameras, and recording astronomical videos is wise to seek to backup their work in places other than solely on the computer's hard disk drive.

Floppy Disks

In the early to mid-1980s, before high-capacity hard drives became widely available, PC programs were stored on floppy disks – thin magnetic disks contained within plastic sleeves. The original floppy disks were a giant 200 mm (8 in.) in diameter, and they really were floppy; the sleeves were made of thin plastic and were capable of being bent to a certain extent, but they became unusable if the bending was taken to extremes. Their capacity ranged from 80 kb to 1.2 MB.

At Yardley's School in Birmingham, UK, we used such gigantic floppy disks to program a large, thoroughly imposing black metallic-cased machine. This author has written several basic ('basic' in both senses of the word) astronomical programs to play with in the lunch break, including a comet orbit program that displayed a change in the length of the comet's tail (well, a single line) according to its distance from the Sun – heady stuff. IT (Information Technology) was then something that students did in their spare time and was not considered important

P. Grego, *Astronomical Cybersketching*, Patrick Moore's Practical Astronomy Series, DOI 10.1007/978-0-387-85351-2_5, © Springer Science+Business Media, LLC 2009

enough to actually teach formally. One program in particular that amused and/or frustrated all its users was called ****luna. The program demanded that the hapless operator land an impossibly fragile lunar module on the surface of an incredibly jagged lunar surface using a limited amount of fuel and controlling thrust and direction with the arrow keys. Although the graphics were mono-chrome and consisted of simple lines, one could imagine the tension experienced by Neil Armstrong as he guided the *Eagle* down to Mare Tranquillitatis (or so we deluded ourselves into thinking).

A somewhat smaller 5.25-in. floppy disk format was also introduced at around the same time, having a capacity of either 360 kb (low-density) or 1.2 MB (high-density). However, an even smaller format, the much more manageable 3.5-in. floppy disk, became the most popular portable storage media of them all. Before the decade of the 1980s was ended, virtually every computer boasted its own 3.5-in. floppy drive. Encased in rigid plastic the disks were hardly floppy, and many purists use the term 'diskette.' Some older computers still have the 3.5-in. floppy drives, but as the diskette's storage capacity ranges from just 720 kb (low density) to 1.44 MB (high density), they can only be used to store small files. Few new computers contain these drives, and those wanting to add one to their system usually opt for a USB floppy drive.

CD/DVD Storage

Other storage media of the 1990s included Iomega's Zip drive (introduced in 1994), which had a storage capacity from 100 to 750 MB, and later the Jaz drive, which boasted a capacity of 1–2 GB. Both formats have been discontinued. Aside from technical problems, two main factors worked against these formats becoming hugely popular – the relatively high price of the disks and the growing availability of writable (and then rewritable) CD-ROMs that, for a fraction of the cost, could store up to 650 or 700 MB of data.

CD-ROMs (compact disk read-only memory) were introduced by the electronic giants Sony and Philips back in 1985, a couple of years after the marketing of the first audio CDs (compact disks). With an identical format to audio CDs, the CD-ROM measures 120 mm (4.75 in.) across and 1.2 mm (0.05 in.) thick; the disk's main body is made from a polycarbonate plastic coated with thin layers of reflective metal (commonly aluminum, but sometimes gold) covered with a protective layer of plastic lacquer. The disks are read with a suitable drive that uses a sophisticated optoelectronic tracking module to direct a low-power laser beam toward the surface of the CD-ROM. As the drive scans a spiral track (which winds outward from the center of the disk to its edge), binary data are read by detecting whether the laser beam is reflected (by an unchanged surface) or scattered (by microscopic pits) along the track. CD-ROM drives have their own speed of data reading; a speed of $1\times$ equates to a data transfer rate of 150 kb per second, while a $52\times$ drive will read up to 7.6 MB per second (from the outside portion of the disk where the speed of data being scanned is the greatest).

One of the things that amazed early users was the durability and apparent resilience of the disks, and their ability to operate even though they were seemingly damaged. Since the data layer is nearest the upper (label) side of the disk, minor scratches and defects on the surface of the lacquered underside are slightly out of focus during the reading process, enabling data to be read from disks in apparently poor condition. In a now famous scene on BBC TV's *Tomorrow's World* a presenter gleefully spread strawberry jam on a CD and managed to play it afterward – something that obviously couldn't have been done with a vinyl disk or an audio cassette.

In 1990, CD-R (CD-Recordable) disks were introduced, allowing computer users to store large quantities of data using a special CD-R drive. CD recorders write data onto CD-Rs using a laser beam, which causes a layer of photosensitive dye in the disk to change color; these color changes are able to be read by a standard CD player (if an audio disk) or CD-ROM drive (audio or data disk).

While CD-R data are generally considered to be a permanent form of storage, with an oft-touted design lifetime of up to a century or more, errors can appear due to a gradual change in the dye's properties, which goes by the decidedly unglamorous term of 'CD rot.' Disk quality varies from brand to brand, and as a general rule cheap disks are liable to degrade the quickest.

CD-RW (CD-ReWritable) disks made their debut in 1997. As their name suggests, data can be written, erased, and rewritten on them numerous times. Unlike CD-Rs, CD-RWs contain a recording layer composed of a phase change alloy of silver and other metals (including indium, antimony, and tellurium), which is heated and melted by a laser beam to produce areas of differing reflectivity; these areas are read just like the data on a regular CD-ROM.

An impressive six-fold increase in data storage – from 700 MB to 4.7 GB – was made possible with the arrival of the DVD-R (Digital Versatile Disk Recordable), which also made its debut in 1997. Although they share the same shape and size, DVD-Rs are more sophisticated than CD-Rs. Their data capacity is greater because they have a smaller pit size and narrower track pitch of the groove guiding the laser beam, and their reading is made possible by virtue of an optical lens with a higher resolution than those found in a CD recorder. The last major advance in optical storage media to affect the general consumer came in 2000 with the introduction of DVD-RW (DVD ReWritable) disks. These work along the same lines as CD-RWs and are currently among the most commonly used mass storage media. A competing format, DVD+RW, is more versatile because the disks can be written onto without erasing the existing data contained on them, as when writing onto a standard DVD-RW.

Flash Memory on the Cards

Before the turn of the twentieth century, most computer users wanting to save their work and transport it to another computer had two portable storage options – the exceedingly humble 1.44 MB capacity of the standard 3.5-in. floppy disk or the much higher capacity (700 MB) but somewhat cumbersome (120 mm diameter) CD-ROMs. The answer to everyone's portable storage headaches came in the form of flash memory, actually invented in 1981 by Fujio Masuoka of Toshiba. This new form of non-volatile solid-state memory chip was patented as EEPROM (electrically erasable, programmable read-only memory) and termed 'flash memory' because the process of deleting the chip's memory contents reminded Masuoka of a camera flash.

Flash memory chips consist of a matrix of rows and columns with two microscopic transistors (separated by a thin oxide layer) positioned at each matrix intersection. One transistor is known as a floating gate, while the other is the control gate. The floating gate has a value of 1 when it is linked to the row via the control gate; however, the value can be changed to 0 through Fowler–Nordheim tunneling (a process far beyond the scope of this modest book to explain in detail). Data are stored digitally, with very fast read access times and low power consumption.

Although Masuoka's invention took a while to catch on, flash memory is now used by a host of electronic devices, including computers, PDAs, cameras, mobile phones, and portable audio/video players. Most of these devices have internal flash memories of varying storage capacities and a slot in which to insert a portable memory card (often with a far larger capacity than the device's built-in flash memory).

During the 1990s most laptop computers (and some handheld computers) were provided with slots to accommodate a PC Card. Originally called a PCMCIA Card (after the Personal Computer Memory Card International Association), they were, as their name suggests, originally intended for the expansion of flash memory (Figure 5.1). There are four types of PC Card (classed according to their thickness); all measure 54 mm wide and all have an identical 68-pin dual-row connecting interface. The PC Card slot can be used for peripheral devices, such as modems and WiFi adapters, and they can also take PC Card flash memory. Most PC Card memory comes in Type I format (3.3 mm thick), but this is not commonly used these days because there are more convenient forms of data storage available. PC Card hard drives are also available in Type III format (10.5 mm thick). PC Cards are used mainly by those with older (but perfectly serviceable) laptop computers. Amateur astronomers are notable users of such vintage technology because they often work in the dark in the field in adverse conditions such as cold and damp and older equipment is less costly than the new ones, making it to some extent expendable.

Many early portable devices, such as digicams and PDAs, possessed an internal flash memory alone; the memory contents could be downloaded onto a computer via a serial or USB (Universal Serial Bus) connection but usually required a device-specific program (provided on a CD-ROM and bundled with the product) to access the memory. For example, a Casio QV-11 (one of the first digicams to deploy an LCD view screen) required a serial connection and the use of Casio's *QV-Link* software. My first experiments in combining digital technology with traditional observing techniques were performed using this camera, and a discussion of these techniques is provided later in this book.

Early PDAs were restricted to onboard flash memory alone, requiring a physical connection to a desktop computer, but within the space of a few years, a host of electronic devices, including PDAs and digital cameras, were rendered far more versatile by the inclusion of a slot for a flash memory card. Flash memory cards are generally robust, capable of withstanding the kinds of jolts and knocks that would put a hard drive out of commission, and they are able to perform in a wide range of temperatures – perfect for taking out and about in the field and carrying in the pocket.

SanDisk introduced the Compact Flash (CF) Card in 1994; measuring 43×36 mm, CF comes in two types – the 3.3-mm thick Type I and the 5-mm thick Type II. The Hewlett-Packard Jornada 540 (made in 2000), for example, came with 16 MB of internal RAM and was provided with the more common CF Type I card slot (Figures 5.2 and 5.3). Compact Flash memory is still going strong; despite its being bigger than most other flash memory cards on the market, it is considered

PCMCIA Card adapter

SD Card

Figure 5.1. The author's Jornada 720 handheld computer, showing the location of the PCMCIA Card adapter (with SD card inserted) – useful for storing and transferring data and graphics files between computers (photo by Peter Grego).

one of the most reliable memory formats and is used in many models of digital cameras and digital SLRs (single lens reflex cameras). As of the time of this writing, CF cards are available in capacities ranging between 512 MB and a staggering 64 GB.

Direct competition with the CF and PCMCIA formats appeared in 1995 with SmartMedia (occasionally referred to as the Solid State Floppy Disk Card, or SSFDC), Toshiba's own flash memory standard. Wafer thin at just 0.76 mm and measuring 45 × 37 mm, their capacity ranged from 0.5 MB to an upper limit of 128 MB. However, SmartMedia memory cards and the devices that use them have long ceased to be produced, largely owing to their inadequate maximum capacity (Figure 5.4).

Several other major electronics companies brought out their own varieties of memory card, notable among them being Sony's Memory Stick, first introduced in 1998, and the Olympus/Fujifilm xD Card, which appeared in 2002. With the size and thickness of a stick of chewing gum, the original Sony Memory Sticks had a capacity ranging from 4 to 128 MB and were compatible with numerous Sony products, including its Cyber-shot range of digital cameras and its Clié series of

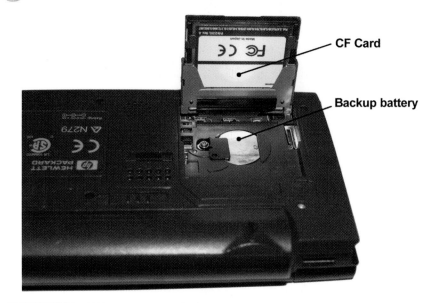

Figure 5.2. The author's Jornada 720 handheld computer, showing the location of the CF Card at the base of the device, in front of the backup battery (photo by Peter Grego).

Figure 5.3. The author's Jornada 540 PDA, showing the location of the CF Card at the top of the device, in front of the backup battery (photo by Peter Grego).

Figure 5.4. A Flash Path adapter, allowing a SmartMedia card to be read from a regular floppy disk drive. SmartMedia is a flash memory format no longer produced (photo by Peter Grego).

PDAs. Sony later abandoned the original large Memory Stick in favor of the smaller Memory Stick Duo, whose format is actually slightly smaller than the standard Secure Digital (SD) Card (see below). xD (eXtreme Digital) Cards are mainly used in digital cameras; they measure $20 \times 25 \times 1.8$ mm and range between 16 MB and 2 GB in capacity.

Introduced in 2000, the Secure Digital (SD) Card has become the most popular type of portable flash memory in use today. SD memory cards were developed by the companies Matsushita, SanDisk, and Toshiba for data storage in portable devices, particularly in digital cameras and PDAs. Measuring a postage stamp-sized 24×32 mm and just 2.1 mm thick, standard SD cards come in a wide range of capacities, from 8 MB to 4 GB. Now SDHC (Secure Digital High Capacity) Cards up to 32 GB are available (Figures 5.5, 5.6 and 5.7). Variants of the SD Card now include the thumbnail-sized SD Mini (20×21.5 mm) and the little fingernail-sized SD Micro (measuring just 11×15 mm). Both of these smaller items can be inserted into standard SD-size adapters for use in devices with regular SD Card slots (Figure 5.8).

Several other types of portable data storage systems are worth mentioning. Microdrives are miniature 1-in. hard drives with capacities of up to 8 GB, designed to fit in a Compact Flash (CF) Type II slot. They tend to require more power than standard CF Cards, making them unusable in many PDAs and other low-power devices; unlike CF Cards they are prone to failure if subjected to sudden jolts. Larger, but still very portable, USB hard drives are available in a variety of forms and capacities, including the Western Digital Passport drive, which measures a pocket-sized 135×85 mm, is just 20 mm thick, and is available up to a capacity of 320 GB. Iomega's Rev drives use removable cartridges containing hard drives with data storage capacities ranging between 35 and 120 GB. They may offer an ideal storage solution in many circumstances, but the relatively high price of each hard drive Rev cartridge in comparison with regular portable USB hard drives makes

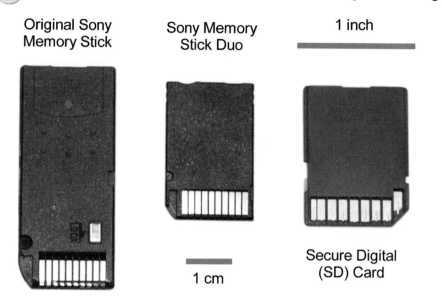

Figure 5.5. An original Sony Memory Stick, Sony Memory Stick Duo, and Secure Digital (SD) Card compared (photo by Peter Grego).

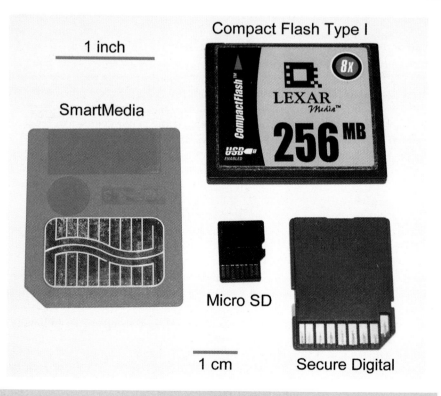

Figure 5.6. Flash memory cards compared (photo by Peter Grego).

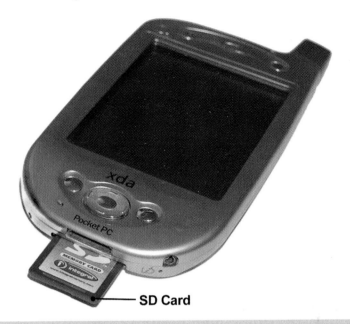

SD Card

Figure 5.7. The author's XDA PDA, showing the location of the SD Card at the base of the device (photo by Peter Grego).

Figure 5.8. A regular SD adapter for an SD Micro Card (photo by Peter Grego).

them somewhat uncompetitive. USB mass storage devices of the flash memory type include lipstick-sized memory dongles; some are just plain looking sticks, but others have a scrolling display listing the contents, and some even have an LCD view screen upon which to browse the contents and view images without having to connect it to the computer.

Faced with such an array of memory cards and portable storage media – big and small, old and new, contemporary and obsolete, plain and cool, low MB capacity to multi GB capacity – the modern computer user is presented with the challenge of being able to successfully read from them on the desktop computer. Although many PDAs and digital cameras are provided with their own USB cable through which a direct computer link can be established, it's often more convenient to read and write data directly from and to the memory card. Most new computers have removed the old 3.5-in. floppy drive and replaced it with a bay containing a complement of USB ports and slots for the most commonly used storage media, such as Compact Flash, Secure Digital, SmartDisk, xD, and Memory Stick. USB adapters are available with slots for virtually every type of memory card; these are particularly useful on laptop and tablet computers, which don't have their own memory card slots (Figures 5.9 and 5.10). Most of these adapters are 'plug and

Figure 5.9. An SD to USB adapter (photo by Peter Grego).

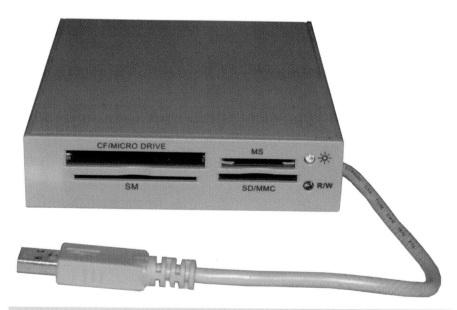

Figure 5.10. A USB flash card reader, capable of reading and writing a variety of memory cards (photo by Peter Grego).

play,' meaning that they don't require special software to install. Instead, the card and its contents shows up under a designated drive letter in the computer's listing of 'devices with removable storage' and can usually be accessed in the same manner as the contents of the computer's hard drives.

Part II

Software and How to Use It

CHAPTER SIX

Electronic Skies

We've already taken a grand tour of computers of various sorts, plus their many peripherals, and assessed their suitability for desktop and field astronomy and cybersketching. In this part we take a closer look at the various kinds of software that bring the skies to our screens and enable us to convert our stylus strokes into cybersketches. A complete guide to the relevant astronomical software and all the graphics programs available for PCs and PDAs will not be presented here. However, we will attempt to present as balanced a view as possible, pointing out a few of the better programs and discussing their merits when applied to cybersketching.

Generally speaking, astronomical software allows the observer to plan and research observations, investigate the circumstances behind a wide variety of celestial objects and phenomena – past, present, and future – and view it in a graphic format. Astronomical software can be used on the desktop and in the field to produce detailed data and accurate graphical templates that assist in the making of observations and astronomical cybersketches.

As we have seen, astronomy – a science that is heavily reliant on a great deal of computational work – benefited considerably from the development of the computer. Amateur astronomy was one of the first hobbies to benefit from the revolution in home computing, which began during the late 1970s. Books such as Peter Duffet-Smith's *Practical Astronomy with your Calculator* (CUP, 1979) gave detailed instructions on how to perform a wide variety of useful astronomical calculations on a calculator or a computer.

P. Grego, *Astronomical Cybersketching*, Patrick Moore's Practical Astronomy Series, DOI 10.1007/978-0-387-85351-2_6, © Springer Science+Business Media, LLC 2009

Many of the first programs for computers were written in BASIC, and they were never intended to produce rich, eye-popping graphical displays. The best of them displayed blocky stars that looked less stellar and more like the Borg cubes of *Star Trek* fame. Amazing graphics were to come later, when computing power had developed to a level sufficient to take the viewer on virtual tours of the Solar System. *Dance of the Planets* by ARC Science Simulations (first released in 1990 and run on DOS) was the first popular program that was able to transform celestial pinpoints of light into recognizable worlds in motion; the program used rendering to present planets in high (VGA) resolution, with accurate phase and surface detail, when these objects were zoomed-in upon. This groundbreaking program also displayed animations of the motions of objects in the Solar System according to various parameters set by the user.

Even die-hard lovers of ancient software would admit that *Dance of the Planets* appears pretty old hat in comparison with the latest astronomy programs. Nowadays there are dozens of full-featured, graphics-intensive astronomical programs available for the PC as downloadable freeware, off-the-shelf or downloadable commercial products or as online programs accessible through the Internet. Although the capabilities of astronomy programs vary considerably, there are several basic categories: planetaria, celestial tourism, subject-specific software, databases, reference works, atlases, peripheral software, and utilities. Most astronomy programs feature combinations of some or all of these functions to a varying extent, but not all of them are relevant or useful for cybersketching purposes. One example is a program such as *Starry Night Pro*, which features a sumptuous planetarium, a 3D mode, has numerous built-in databases and reference material, and can be used for telescope control.

Planetaria

Planetarium programs are among the most useful, delightful, and entertaining of all astronomy computer programs – and not only amateur astronomers. They present a graphic display of the celestial sphere based on certain user-defined parameters, such as date (which, in many programs, covers a range of many thousands of years before and after the present), time, and geographical location. Depending on the individual program's capabilities, objects displayed on the celestial sphere can include some or all of the following (to name but a few):

Celestial sphere – shown in equatorial projection, with RA and Dec labels, or presented as an altazimuth, ecliptic, or galactic grid projection. Local horizon is also displayed.

Stars – millions of them, from the Hipparcos and Tycho databases, displayed according to magnitude, with size, brightness, and labels adjustable by user. Double and variable stars are shown. Constellations, their connecting lines, and boundaries can be displayed, as well as well-known asterisms.

Planets – planetary positions, displayed as points, labels, or icons, with a tracking facility to show their motions against the celestial sphere over time. Planets can be zoomed-in to show their phase and features, even overlain with reference grids to show their orientation. Planetary satellites are also displayed. Dwarf planets and asteroids are also shown. Data on each of these objects can be displayed on-screen.

Moon – an accurate representation of the main features of the lunar surface, with nomenclature. The lunar disc is displayed with the correct phase and libration. Solar and lunar eclipses and occultations can be simulated.

Comets – the positions of periodic comets and brighter non-periodic comets can be shown. The user can add data (manually input or automatically updated) to include newly discovered comets.

Deep sky objects – in addition to the 110 Messier objects, thousands of NGC and IC objects, many tens of thousands of objects in deep space – galaxies, globular clusters, supernova remnants, planetary nebulae, reflection and emission nebulae, and more – can be located.

Celestial Tourism

This type of astronomy program allows the user to move around a virtual 3D space environment to view objects from different angles. Some programs, such as RITI's *Celestial Explorer Mars,* enable planetary surfaces to be mapped, zoomed-in upon, and viewed in 3D 'flyaround' mode, while other programs take the user further afield, into roamable stellar and intergalactic realms, notable among these being *Deep Space Explorer* by Space.com.

Programs dedicated to one particular subject, such as Jupiter (*JUPOS*), lunar occultations (*Lunar Occultations Workbench*) or the Moon (Virtual Moon Atlas), often contain more detailed information and more observationally oriented graphics than full-featured planetaria. Much of this specialist software is written by people for the love of the subject, and many of these programs (including the three mentioned above) are freeware.

Databases and Other Reference

Many databases are accessible online for free (although some require registration), for example *The Deep Sky Database* at http://www.virtualcolony.com/sac/. There are also innumerable online and commercial sources of astronomical reference. To name a few examples among the thousands of excellent web-based resources is *Eric Weisstein's World of Astronomy* at http://scienceworld. wolfram.com/astronomy/, *Astronomy Online* at http://astronomyonline.org, and the Astronomical Data Center at http://adc.gsfc.nasa.gov/. *Wikipedia* at http://www.wikipedia.org/ also contains a great (and ever-greater) amount of astronomical information. Specialist material inclined toward observation includes the *Telescopic Companion to the Finest NGCs* and the *Telescopic Companion to the Messier Objects* by Roger Fell, available from http://www.starastronomy.org/Observing/Fell/. Commercial reference suites on CD and DVD-ROM include Dorling Kindersley's *Encyclopedia of Space and the Universe.*

Peripheral Software and Utilities

Astronomy-related peripheral software includes items such as Meade's *Autostar* and Celestron's *NexRemote* telescope control suites, programs for adjusting a monitor's gamma and color to facilitate field observing in dark-adapted mode, GPS software, and CCD imaging software. Utilities of practical use include accurate time displays (both UT and sidereal) for precise timing work, eyepiece calculators, and online weather resources.

Astronomy Programs for Planning and Researching Observations

Here is a selection of astronomy programs useful for planning and researching observations.

Platform (various OS versions): Win – Windows, Mac – Apple Mac, PPC – Pocket PC, Palm – Palm PDA
Availability: F – Freeware, S – Shareware, C – Commercial product

Title	Platform	Availability
Astromist	PPC/Palm	C

http://www.astromist.com/

A great piece of astronomical software for Palm and P/PC, *Astromist* is highly versatile and very powerful planetarium program for the PDA, with support for GPS and wireless telescope control (Figure 6.1). It makes a superb field companion.

CalSky	Internet	F

http://www.calsky.com/

Founded in 1991, *CalSky* is a worldwide interactive online astronomical/space calendar and calculator, intended for friends of astronomy as well as astronomers. There's a huge range of tools to choose from – many of them capable of being tailored to the observer's site and set up – to plan observing sessions. In addition, there's a wealth of background detail. And it's all free, accessible online through a desktop PC or through a WiFi-enabled computer in the field.

Cartes du Ciel	Win	F

http://www.stargazing.net/astropc/index.html

An excellent, full-featured freeware planetarium and sky mapping program (Figure 6.2).

Celestia	Win/Mac/Linux	F

http://www.shatters.net/celestia/

22/08/2008 03:45

Maurolycus

Figure 6.1. Screenshot from the P/PC version of *Astromist* showing a view of the Moon with feature identification.

Celestia's superb graphics lets the user explore the universe in three dimensions. This top-notch freeware program can be used to travel through the Solar System, to any of over 100,000 stars, or even beyond the galaxy. Planetary images brought up in the program can be modified and used as cybersketching templates (Figure 6.3).

Celestial Explorer Mars	Win	C

Celestial Explorer Mars by RITI allows the user to explore the fascinating Martian landscape through real-time 3D modeling and visualizations based on

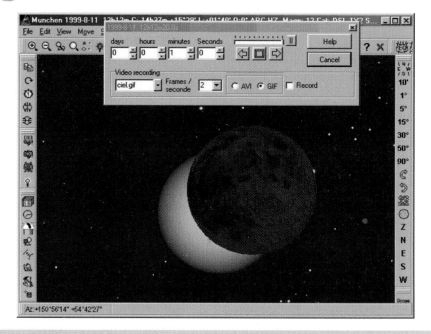

Figure 6.2. Screenshot of a solar eclipse simulation in Cartes du Ciel.

Figure 6.3 Screenshot from Celestia, showing a close-up of Saturn.

the latest high-definition maps of the red planet. It's fascinating and educational (even fun) to use. Moreover, aspects of this program are also useful for planning observations and preparing observing blanks for cybersketching, since the settings can be adjusted to display a full disk of Mars, showing its features and its phase and tilt as viewed telescopically from the Earth.

DarkAdapted	Win/Mac	F

http://www.aquiladigital.us/darkadapted/index.html

This nifty program enables the monitor's gamma and color settings to be modified to best preserve night vision. It's ideal for use in the field with a laptop or tablet PC while observing, imaging, or cybersketching – the system-wide menu, accessible through assigned hot keys, gives full control over *DarkAdapted* from within any application (Figure 6.4).

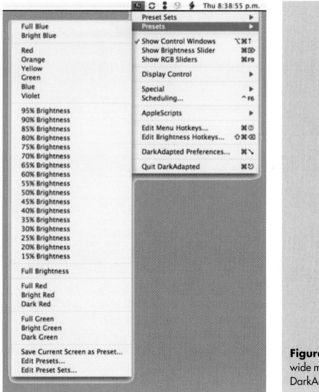

Figure 6.4. System-wide menu of DarkAdapted.

Deepsky	Win	C

http://www.deepsky2000.net/index.htm

A superb program enabling detailed planning of astronomical observations – useful not only for deep sky observation, but also for lunar, planetary, and other Solar System work. The user's own observing notes, images, and cybersketches can be logged and viewed in a personal observing journal (Figure 6.5). Its sister program for P/PC, *Pocket Deepsky* (below) comes free with the full Windows program.

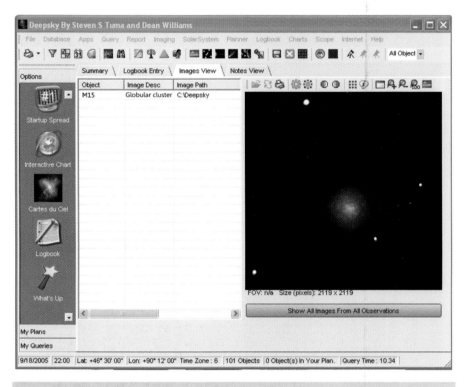

Figure 6.5. Screenshot of Deepsky, showing a cybersketch of M15 as a journal entry.

Deep-Sky Planner   Win   C

http://knightware.biz/dsp/

Another superb program enabling desktop and field observation planning and logging, not only for deep sky observation but for planets, comets, and minor planets, too. The observing log in *Deep-Sky Planner* is electronically portable, either as HTML or text reports or as import/export files. The log management feature helps you to select observations for exporting, and conversely, files for importing – a useful feature for transferring images and cybersketches.

Distant Suns   Win/Mac   F/C

http://www.distantsuns.com

One of the longest established astronomy programs, *Distant Suns* provides a versatile tool for the amateur astronomer.

Earth Centered Universe   **Win**   **C**

http://www.nova-astro.com/

A planetarium and telescope control program.

Encyclopedia Galactica   **Win**   **F**

http://www.eureka.ya.com/javastro/index.htm

A freeware planetarium program that allows the creation of different types of sky maps in which practically any element is customizable. In addition to this basic functionality, it also provides a number of tools, among them calculating the positions of the moons of Jupiter and Saturn, a Moon calendar and feature locator, an observing list manager, and maintaining a customizable database of celestial object images (Figure 6.6).

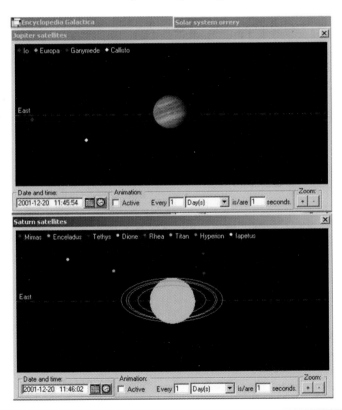

Figure 6.6. Screenshot from *Encyclopedia Galactica* showing Jupiter and Saturn and their satellites.

GeoAstro Applet Collection	Java Applets	C

An incredible collection of Java applets presenting valuable data on a number of astronomy-related themes.

GrayStel Star Atlas	Win	C

http://members.aol.com/graystel/gsahp.htm

An excellent program that combines all the accuracy of a computer-generated star atlas with seamless photographic mapping. The Moon is displayed photographically with images taken at different phases, while planetary views depict accurate phases and tilts, useful for cybersketching template preparation (Figure 6.7).

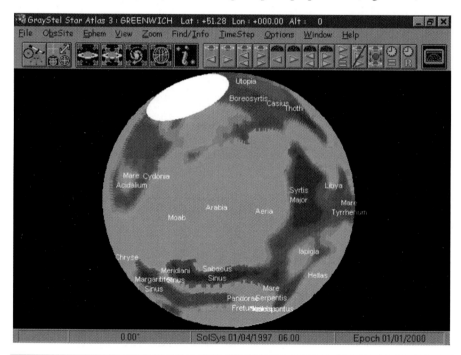

Figure 6.7. Screenshot from *GrayStel Star Atlas* showing a close-up of Mars.

Google Earth	Internet	F/C

http://earth.google.com/

An amazing program requiring an initial software download and Internet access, *Google Earth* lets you fly anywhere on Earth to view satellite imagery, maps, terrain, and 3D buildings; it also has an interactive planetarium program so that the night sky can be explored in detail. Downloading the basic version and subsequent access is free.

HNSky (Hallo Northern Sky) Win F

http://www.hnsky.org/software.htm

A very capable planetarium program, *HNSky* is complete with the SAO, PPM, Tycho-2 star databases, an up-to-date deep sky database containing 26,000 objects, Realsky images, and access to the external GSC and USNO CD-ROM star catalogs. Telescope control through the program is possible using the universal ASCOM package (Figure 6.8).

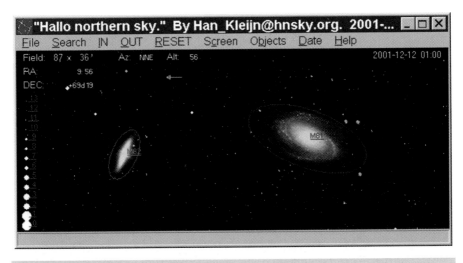

Figure 6.8. Screenshot of galaxies M81 and M82 using GSC CD-ROM database and provided FITS images, from HNSky.

Home Planet Win F

http://www.fourmilab.ch/homeplanet/

A very good freeware planetarium program with clear charts – but beware the cuckoo clock! Available in several versions, including a screen saver.

JUPOS Win F

http://jupos.privat.t-online.de/

A specialized program for the dedicated observer of Jupiter and its satellites. *JUPOS* is designed for amateur astronomers to record and analyze positions of features on Jupiter and to display them in suitable time–longitude or other coordinate systems. Images and visual observations (including cybersketches) can be processed and measured, enabling feature positions to be displayed graphically. Cybersketch templates for Jupiter can be prepared from the program's graphic displays.

K3CCDTools   **Win**   **F/C**

http://www.pk3.org/Astro/

Highly capable astro-imaging software that includes video capturing, frame aligning, and stacking/summing. It also provides simple image post-processing.

Lunar Map Pro   **Win**   **C**

http://www.riti.com/prodserv_lunarmappro.htm

A comprehensive lunar mapping program for the amateur astronomers. *Lunar Map Pro* offers high-resolution maps, corrected in real time, showing accurate phase and libration corrections. The maps can be customized to match the view through your own instrument, and they can be used as templates for lunar cybersketches.

Meridian   **Win**   **F**

http://www.merid.cam.org/meridian/english.html

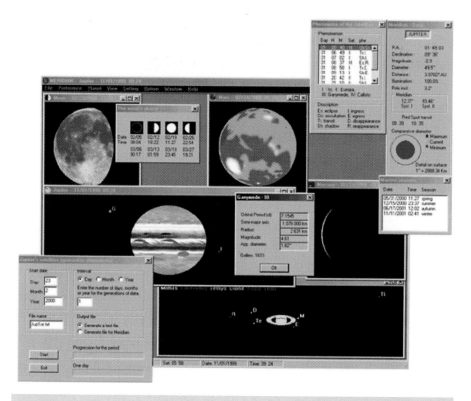

Figure 6.9. Screenshot from Meridian.

A relatively simple program but very useful for the Solar System observer, with no-nonsense views of the planets, plus salient information, shown for any time and date requested. *Meridian* is good for researching and planning observations and preparing cybersketches (Figure 6.9).

NGCView	Win	S

http://www.rainman-soft.com/

NGCView enables astronomical observations to be planned and logged with the aid of a thorough database and star atlas.

Pocket Deepsky	PPC	C

http://www.deepsky2000.net/pocket.htm

Figure 6.10. PDA screenshot from Pocket Deepsky.

Like its stable mate, the Windows program *Deepsky* (above), *Pocket Deepsky* enables detailed planning of astronomical observations – useful not only for deep sky observation but also for lunar, planetary and other Solar System work. The user's own observing notes, images, and cybersketches of objects can be logged and viewed as a personal observing journal (Figure 6.10).

Pocket Stars   **Win/PPC**   **C**

http://www.nomadelectronics.com/

An accurate star chart and guide to the night sky for PC and Pocket PC.

Redshift   **Win**   **C**

http://www.redshift.de/gb/_main/index.htm

A long-established, full-featured planetarium program featuring an extensive catalog of stars and deep sky objects, most useful for planning observations and research. Features a handy sky diary (Figure 6.11).

Figure 6.11. Jupiter and its moons, displayed by Redshift 5.

Registax   **Win/Linux**   **F/C**

http://www.astronomie.be/registax/

One of the most capable tools for aligning, stacking, and processing BMP or AVI image sequences to produce enhanced images of astronomical objects (Figure 6.12).

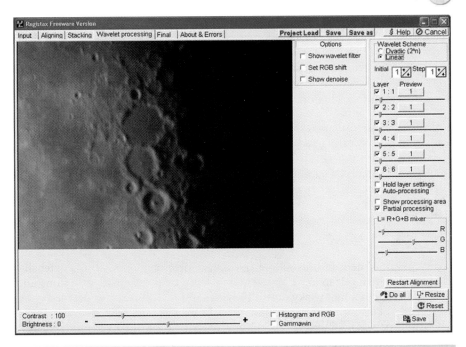

Figure 6.12. Digicam image of lunar craters being processed in Registax.

Sky & Telescope Observing Tools	Internet	F

http://www.skyandtelescope.com/

A suite of web-based tools for the amateur astronomer, including an interactive sky chart, almanac, lunar phases, Mars profiler, Jupiter's Red Spot and moons, the moons of Saturn and Uranus, and Triton Tracker. Online registration is required, but subsequent use is free.

SkyChart III	Win/Mac	S

http://www.southernstars.com/skychart/index.html

An advanced planetarium program that accurately simulates and displays the sky, past, present, or future. *SkyChart III* supports a number of popular computer-controlled telescopes and digital encoders, enabling computer-aided observing. The database contains the entire *Hubble Guide Star Catalog* (over 19 million objects) and a smaller database of 300,000 objects based on NASA's *SKY2000 Master Star Catalog*.

SkyMap Pro	Win	C

http://www.skymap.com/

An excellent planetarium and star-charting program for the serious amateur astronomer.

Starry Night   Win/Mac   C

http://www.starrynightstore.com/

A richly featured planetarium program with one of the most pleasant graphic displays of all. Featuring more than 19 million celestial objects, *Starry Night Pro* offers instantaneous access to a 3D implementation of the complete Hipparcos star database, plus the Hubble GSC, NGC/IC, and PGC catalogs, as well as the Messier deep-space objects. User images (including cybersketches) can be added to the deep sky database.

Taiyoukei   PPC   F

http://freewareppc.com/astronomy/taiyoukei.shtml

An excellent freeware program with several add-ons. It features detailed ephemerides of Solar System objects, chart views of planets and their moons, eclipses, meteor shower data, a celestial object monitor (with altitude and azimuth gauges), and a great deal more besides.

The Caldwell Observing Log   Win/Mac   F

http://www.davidpaulgreen.com/tcol.html

The Caldwell Observing Log describes the 109 objects described by Sir Patrick Moore in his Caldwell Catalog. It features location charts complete with spaces for inputting the relevant observational data; observational notes, images, and cybersketches can be inserted into the digital forms provided. A good cybersketching feature allows any graphics program to be called up via the program's interface (Figure 6.13).

TheSky   Win/Mac/PPC   C

http://www.bisque.com/products/thesky6/

With one of the best pedigrees in astronomical software, *TheSky* is an easy-to-use, full-featured graphical planetarium program available for a variety of platforms (Figures 6.14 and 6.15).

The Simple Observing Log   Win/Mac   F

This freeware program allows the amateur astronomer to log and keep track of observing information. It features easily completed forms, sortable by date or by object, and there's a space where observational cybersketches can be inserted. An

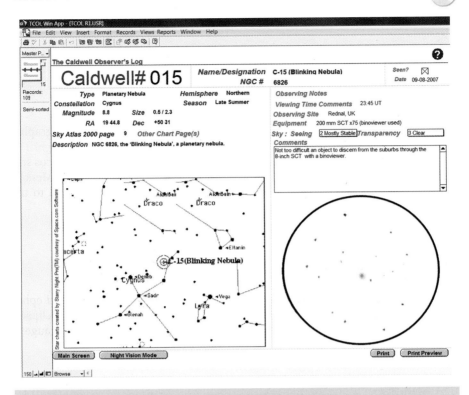

Figure 6.13. TCOL screenshot showing a cybersketch of NGC 6826.

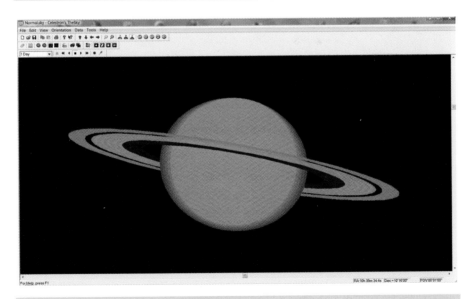

Figure 6.14. Screenshot showing a close-up of Saturn in *TheSky*. Images such as these can be used as cybersketching templates.

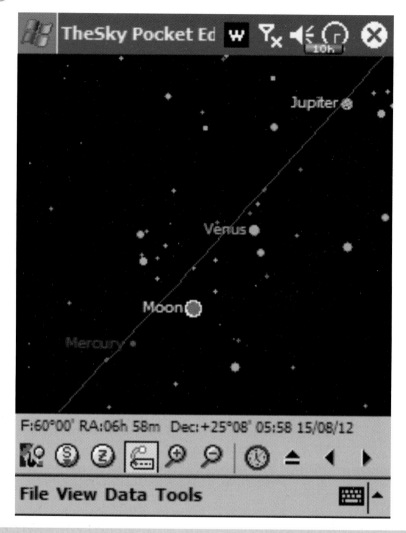

Figure 6.15. Screenshot from the Pocket PC version of *TheSky*, showing the Moon and several planets.

'astrophoto' section for each observation lets astrophotographers keep track of their images as well (Figure 6.16). TSOL can also be used in the field in a red night vision mode.

The Ultimate Messier Object Log   **Win/Mac/Palm**   **F**

http://www.davidpaulgreen.com/tcol.html

From the same stable as *The Caldwell Observing Log* and *The Simple Observing Log* (above), *The Ultimate Messier Log* is a FileMaker Pro 6.0 database containing

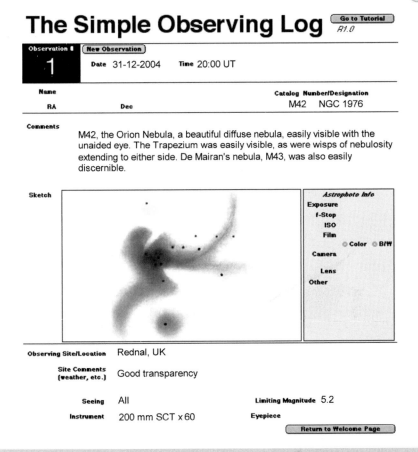

The Simple Observing Log

Go to Tutorial
R1.0

Observation #	New Observation
1	Date 31-12-2004 Time 20:00 UT

Name

Catalog Number/Designation
M42   NGC 1976

RA   Dec

Comments

M42, the Orion Nebula, a beautiful diffuse nebula, easily visible with the unaided eye. The Trapezium was easily visible, as were wisps of nebulosity extending to either side. De Mairan's nebula, M43, was also easily discernible.

Sketch

Astrophoto Info
Exposure
f-Stop
ISO
Film
○ Color ○ B/W
Camera
Lens
Other

Observing Site/Location Rednal, UK

Site Comments
(weather, etc.) Good transparency

Seeing All   Limiting Magnitude 5.2

Instrument 200 mm SCT x 60   Eyepiece

Return to Welcome Page

Figure 6.16. Screenshot from TSOL, showing an observation of M42, the Orion Nebula.

all the information required to observe all 110 Messier deep sky objects. It features space for notes and viewing comments, descriptions, and a special sketch view, useful for inserting cybersketches (Figure 6.17).

Virtual Atlas of the Moon   **Win**   **F**

http://www.ap-i.net/avl/en/download

Virtual Moon Atlas is a great freeware program that delivers a lot of accurate information about the Moon. It has a libration-corrected graphic display and features a range of functions useful to the lunar observer. Zoomed-in portions of the charts may be saved and used as basic cybersketching templates (Figure 6.18).

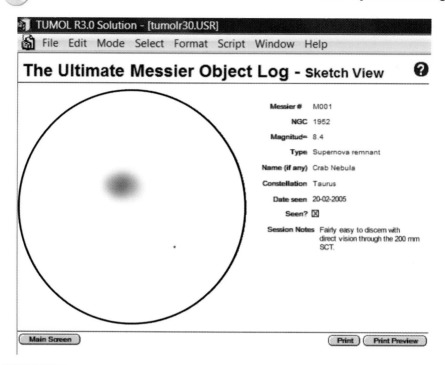

Figure 6.17. TUMOL screenshot showing a cybersketch of M1, the Crab Nebula.

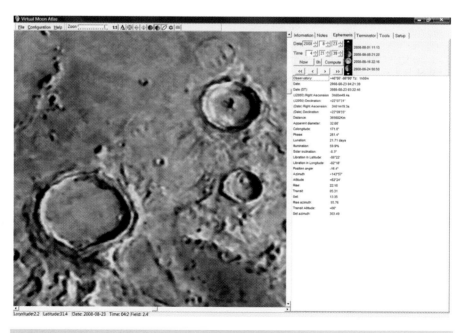

Figure 6.18. This screenshot from *Virtual Moon Atlas* shows a close-up of the craters Aristarchus, Aristillus, and Autolycus. The image may be used as a cybersketch template.

A selection of other useful astronomy programs:

Title	Platform	Availability
AlphaCentaure	Win	S
AIP for Windows (AIP4Win)	Win	C
Astra Image	Win	S
Astro (in French)	DOS	F
AstroByte Logging Software	Win	F
AstroCal	Mac	S
AstroCalc	Win	C
AstroClk	Win	F
AstroGrav	Win/Mac	C
AstroHelper	Win	F
Astro-Mania	Win	S
AstroMB – computer-aided astronomy	Win	C
Astrometrica	Win	S
Astronom	Win	S
Astro Timer	PPC	F
MacAstronomica	Mac	S
WinAstronomica	Win	S
Astronomica for PC	Win	S
L'Astronomy par les calculs	Win	S
Astronomy Calculator	Win	C
Astronomy Explorer	Win/Mac	C
Astronomy Slide Show	Win	F
AstroPlanner	Win/Mac	F
AstroSnap	Win	S
AstroStack	Win	F
AstroVideo	Win	S
AstroViewer	Java Applets	F
Atlas Du Ciel	Win	C
Atlas of the Solar System	Win	C
AudeLA	Win/Linux	F
C88	Win	S
Celestial Maps	Win	C
ChView	Win	S
Clear Skies!	Mac	S
Coeli – AdAstra	Win	S
Coeli – Stella 2000	Win	S
Constellation Star Maps	Win	F
CoolSky	Win	S
CosmoSaver	Mac	S
CyberSky	Win	S
Dance of the Planets	DOS	C
Deep Space	Win	S
Deep Sky Database	Internet	F
DeskNite	Win	F
Desktop Universe	Win	C
The Digital Universe	Mac	C
EasySky	Win	C
Ephemeris	PPC	F

(continued)

Title	Platform	Availability
Guide (Project Pluto)	Win	C
Hipparchus	Mac	S
Hitchhiker 2000	Linux/Unix S	
IrisWin		F
JIMSAIP Astronomical Image Processing	DOS	S
JupView	PPC	F
Lunar Occultation Workbench	Win	F
MacStronomy	Mac	S
MaxClock	DOS/Win F	
Maxlm DL	Win	C
Megastar	Win	C
MoonClock	Mac	S
MoonPhase	Win	S
Moonrise	Win	S
MPj Astro (EquinoX)	Mac	S
MyStars!	Win	S
Night Sky	Mac	S
Night Sky Interactive	Win/Mac S	
Our Cosmohood	Win	S
OrionicPPC		F
PC Sky	Win	C
Perseus	Win	C
Planetarium	Palm	C
Power Age Sky SimulatorWin	S	
SatWin		F
SATSAT Satellites of SaturnDOS		F
Sidereal Clock	Win	F
Sky Screen Saver	Win	F
Sky Sentinel PlanetariumMac	C	
SkyTools	Win	C
Solar Kingdom	Win	S
SpicaWin		F
StarCalc	Win	F
StarCat	Win/Linux S	
StarDrive	Win	S
Stargazer's Delight	Mac	C
Stars of the Night Sky	Win	S
StarStrider	Win	S
Stellaris	Win	S
Stellarium	Win/Mac/Linux	F
The Fourth Day	Win	S
The Planets	Internet (Java)	F
Voyager 4.5	Win/Mac C	
WinOccult	Win	F
XEphem	UNIX/Win/Mac OS	C

PDA Program Installation in Vista

PDA users running a pre-Windows Vista desktop PC will be familiar with the program *ActiveSync* for H/PC and P/PC devices or *HotSync* for Palm devices. These programs allow a link to be established (usually via USB cable) between the PDA and PC, enabling mutual programs to be synchronized, information shared, and files freely moved between the linked computers. If you choose to upgrade your PC to Windows Vista you will find that *ActiveSync* does not work, nor is Palm *HotSync* supported.

ActiveSync is used when attempting to install programs onto a Windows Mobile device using an installation program run from the PC, and it is the only way of installing some software onto the mobile device. There are other means of downloading programs and installing them onto a mobile device. The most commonly used is to simply transfer the relevant CAB file (and other supporting files if necessary) to the main memory or storage card of the PDA and run it from there. Full communications between your Windows Mobile device and a Windows Vista PC can be established with *Windows Mobile Device Center 6.1*, which can be downloaded from http://www.microsoft.com/windowsmobile/en-us/downloads/ microsoft/device-center.mspx. Supported devices include Windows Mobile 2003, Windows Mobile 2003 Second Edition, Windows Mobile 5.0, Windows Mobile 6.1, and Windows Embedded CE 6.0. Older Windows CE devices are not supported, however – exactly why this might be is unclear. Thus, the Jornada 540 is, as a result, unable to communicate with a Windows Vista desktop PC, for example.

Palm PDA users wishing to establish a PC connection will discover that the older versions of Palm's *Quick Install* software does not work on Windows Vista, meaning that Palm programs can't be downloaded through the USB-connected *HotSync* manager to your Palm PDA. Thankfully, there is a workaround. You will need to download and install *Palm Desktop 6.2* and *HotSync Manager 7.0* from the Palm web site at http://www.palm.com/us/support/downloads/windesk62.html. These programs can be installed on Windows Vista (Basic, Home Premium, Business, and Ultimate 32-bit OS) and Windows XP (Home and Professional 32-bit OS), and its integrated desktop is compatible with all Palm OS 3.5.× through 5.4.× devices.

CHAPTER SEVEN

In Graphic Realms

Vectors and Bitmaps

Cybersketching is entirely dependent upon graphics programs, applications that enable images to be created and modified on a computer. There are two basic types of computer graphics – vector graphics and bitmaps. Vector graphics are best suited to computer-aided design, producing technical drawings and designs featuring illustrations with sharp-edged geometrical patterns (Figure 7.1). As mathematically based constructs that use combinations of points, lines, curves, and polygons, vector graphics are capable of being scaled in size without any loss of detail; they are also economical in the amount of computer memory they occupy. Vector images are easily converted to bitmaps (see below).

Bitmap images (also known as rastered graphics) are far more suitable for cybersketching purposes (Figure 7.2). They are composed of a grid of pixels, each of which has a designated color. Two factors determine the quality of a bitmap – its resolution (its height and width in pixels) and its color depth. It may be worth pointing out that it's easy to get confused by the word 'pixel' when talking about images. Although a computer monitor's display is made up of tiny, barely perceptible pixels, don't get the impression that each pixel making up an image can only be displayed by just one screen pixel. Graphics programs make it possible to zoom in on an image so that individual pixels are clearly visible, allowing the user to edit an image minutely.

P. Grego, *Astronomical Cybersketching*, Patrick Moore's Practical Astronomy Series, DOI 10.1007/978-0-387-85351-2_7, © Springer Science+Business Media, LLC 2009

Figure 7.1. The Saturn illustration in the top left corner is a vector image. When zoomed-in, the image maintains its integrity.

Figure 7.2. The Saturn illustration in the top left corner is an RGB bitmap image. When zoomed-in, the image breaks down into individual pixels. The color of each pixel is determined by adding its *red, green,* and *blue* values.

Color Depth

Each pixel making up a bitmap image has its own color value (Figure 7.3). The simplest bitmap images with the minimum possible information content are termed 1-bit monochrome images; each pixel in such an image is given a value of either 0 or 1 (1 bit equals 1 binary digit), so each bit represents two colors. In monochrome black and white images a value of 0 represents black and a value of 1 represents white; similarly, a color can be assigned to either 0 or 1 in the case of a 1-bit monochrome color image. For cybersketching purposes, it is possible to produce work in 1-bit monochrome format – traditional astronomical observers have been doing this sort of thing for centuries by producing line drawings and stippled observational sketches to convey the illusion of different shades of gray (Figure 7.4).

As the number of bits per pixel is increased, the greater the depth of color is displayed. The 8-bit images represent 256 colors – 2 multiplied by itself 8 times $(2 \times 2 \times 2 \times 2 \times 2 \times 2 \times 2 \times 2 = 256)$. The 24-bit images have a color depth of 16.8 million colors – 2 multiplied by itself 24 times $(2 \times 2 = 16,777,216)$. The 24-bit images actually have a greater color range than that accurately discernable with the unaided human eye. The deepest color images are those comprising 32- and 64-bits; such images are useful for scientific and archival work.

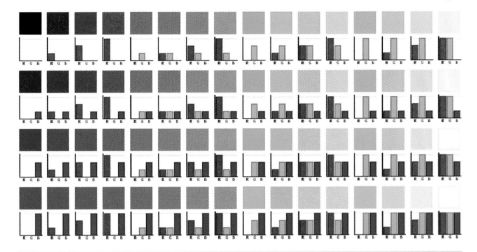

Figure 7.3. Different *red, green,* and *blue color* values assigned to each pixel produce a spectrum of colors.

Figure 7.4. An observational drawing of the lunar crater Lansberg, converted to a 1-bit black-and-white halftone and 8-bit gray scale. The close spacing of black-and-white pixels in the 1-bit image conveys the impression of gray shading.

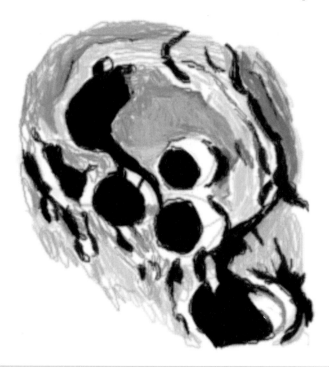

Figure 7.5. Unretouched observational cybersketch of the lunar crater Heinsius, made on a Jornada 540 H/PC using the program *Mobile Atelier* (photo by Peter Grego).

Naturally, the depth of color in an observational cybersketch depends on the computer upon which it is made and the program being used. Basic cybersketching is entirely possible on old PDAs with 1-bit monochrome screens. For cybersketching purposes it is best to work with 8-bit images at the eyepiece; greater color depth may be added when image enhancements are applied on the desktop computer. The Hewlett-Packard Jornada 540 H/PC, for example, has a 12-bit (4,000 color) 240 × 320 pixel screen, and the cybersketching program of choice for this PDA is the very capable *Mobile Atelier,* which saves images as 8-bit BMP files (Figure 7.5).

Color Perception

Now, it must be admitted that generally speaking observational astronomy is not as visually colorful a pursuit as, say, philately or botany. But this is not to say that celestial objects are dull and chromatically challenged – as any glance at the spectacular multi-colored images from the Hubble Space Telescope will so clearly demonstrate.

To help explain things, let's take a look at the way the human eye works. Two types of light-sensitive cells are found in the retina. Cone cells at the retina's center give us detailed color vision at the center of our field view, but these cells are only triggered in bright conditions. When observing the Sun, Moon, and brighter planets, color is clearly visible. Through the eyepiece, the lunar landscape displays here and there some subtle hues of orange, green, and blue. Jupiter's cloud belts show a delightful range of creams, browns, and reds, and the dusky deserts of Mars, though predominantly orange, can display some distinctly greenish tones. Colors on the Moon and planets are enhanced by the use of colored filters.

Only the rod cells around the cones, some distance from the center of the field of view, are triggered in dark conditions. The rods have less resolving power than the cones, so they deliver less-detailed images – and, importantly, the rods cannot distinguish colors. A dim object, such as a diffuse nebula, may be difficult or impossible to see when it is looked at directly. Using a technique called averted vision, observers shift their view a little way to the side of the actual position of the faint object in order to see it better, but it will not have any discernable color. An observer with healthy eyesight may be able to see objects more than two magnitudes fainter using averted vision than with direct vision.

Traditionally, most observers have chosen to depict their subjects in regular pencil grayscales – this is in part due to the fact that colors are often subtle and difficult to distinguish, but also because it is quite obviously rather difficult to make colored pencil drawings (or renditions in pastels or paint) while seated in the dark at the telescope eyepiece. Thankfully, new technology has come to the observer's aid. Cybersketching with a portable computer allows colors to be added to an observational sketch almost as easily as it is to make a grayscale drawing (Figure 7.6).

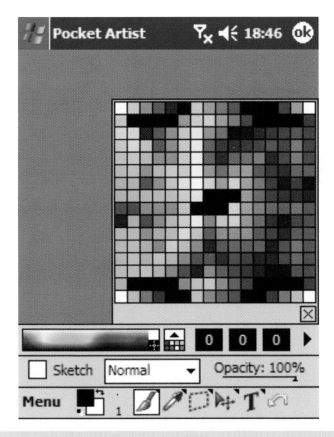

Figure 7.6. The color palette offered by the P/PC program *Pocket Artist 3* (photo by Peter Grego).

Image Formats

Graphics files can take up enormous amounts of disk space. For example, a single raw (unprocessed) image taken with an entry-level 8 megapixel DSLR camera may occupy more than 13 MB of storage space in its original uncompressed format. Therefore, large, high color image files usually require some form of digital compression to squeeze the vital image data to occupy as small a memory footprint as possible. Compression can be either lossless, which preserves all of the image information without loss of picture quality, or lossy, which has the advantage of producing much smaller files but at the same time information is lost each time the file is saved.

Lossy formats are particularly suitable for storing images with large areas of the same color – a lunar image, for example, containing broad areas of shadow. For example, picture a block of color comprising ten rows of ten black pixels.

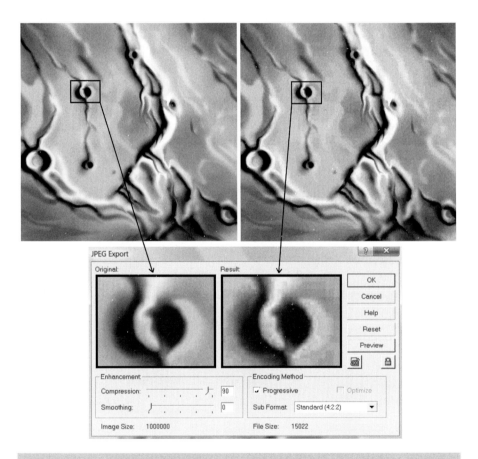

Figure 7.7. The figure shows the degradation in the quality of a 1500 × 1500 pixel 300 dpi grayscale JPEG image of a lunar cybersketch saved with 95% compression (no smoothing) in *Photopaint* (photo by Peter Grego).

Whereas a lossless format will register the value of each individual pixel, a lossy format simply registers the color of one pixel and notes that there are ten rows of ten identical pixels, producing a magnitude's difference in the ensuing file size.

Compression attempts to digitally encode an image while preserving a desirable level of image quality. The amount of compression, and therefore the end result, can be finely controlled by the user. When saving lossy JPEG files, the user is presented with a pop-up dialog screen showing various compression options; in most graphics programs the options also feature preview panes showing the image before and after compression so that the user is aware of the consequences of applying compression before saving the file (Figure 7.7).

Over the years a bewildering number of file formats has sprung up – some of them proprietary, unique to the program in which the graphic was created, and others generic, the most common ones being BMP, GIF, JPEG, PNG, and TIFF. Different file formats have different color depth capabilities. Graphics programs for personal computers usually present a range of image-saving options, from color depth to file format, but most PDA graphics programs save graphics in lossy JPG or BMP format by default to save disk space.

Some Commonly Used Graphic Formats

Format	Bitmap/Vector	Color depth (bits)
AI	Vector	(native to *Adobe Illustrator*)
CDR	Bitmap and vector	1–32 (native to *CorelDRAW*)
GIF	Bitmap	1–8
JPEG	Bitmap	8-bit (grayscale), 12–24 color
JPEG 2000	Bitmap	8, 16 (grayscale), <48-bit color
PCX	Bitmap	1–24
PICT	Bitmap and vector	1–32
PNG	Bitmap	1–64
SVG	Vector	24–32
TGA	Bitmap	1–32
TIFF	Bitmap and vector	1–32
XAML	Vector	32–64
BMP	Bitmap	1–32

Images with a limited color palette, such as GIF files, can be dithered to simulate greater color depth. Dithering is a process similar to the way in which the halftone printing technique works, by juxtaposing different colored pixels to produce a range of other colors when the image is viewed at a distance. Conversely, dithering is also used to reduce the color depth of images to produce smaller file sizes, say when converting a large 24-bit TIF image into a compact, web-ready 8-bit GIF image. Dithered images do have their limitations, and highly dithered images tend to display a characteristic grainy, blocky, or speckled appearance (Figure 7.8).

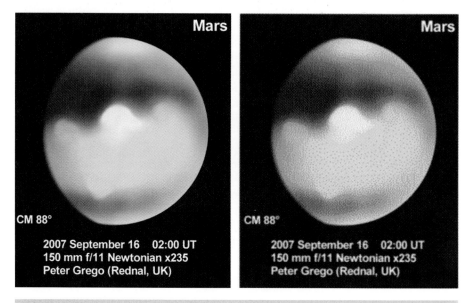

Figure 7.8. The difference in image quality is strikingly evident in this comparison between a 24-bit TIF image (300 dpi, 700 × 875 pixels, file size 1.8 MB), which has been converted to a web-ready 8-bit GIF image (72 dpi, 350 × 437 pixels, file size 32 Kb) (photo by Peter Grego).

Some PDA drawing programs such as *Mobile Atelier* save images in 8-bit BMP format at a resolution of 240 × 320 pixels. This is okay for general astronomical cybersketching at the eyepiece, which doesn't require a great amount of really fine detail. Given the choice, however, it is probably better to work with 24-bit JPEG images in the field and at the desktop. One good handheld cybersketching program is the excellent *Pocket Artist 3*. This versatile program can be run on both the Jornada 720 handheld PC with its letterbox-format screen (with Windows for Handheld PC 2000 OS) and also on a number of pocket PCs, including SPV M2000 (Windows Mobile 2003 Second Edition OS).

When making observational cybersketches on tablet PC from scratch you can use an A6-sized (10.5 × 14.8 cm/4.1 × 5.8 in.) plain white-background template at 300 dpi; choose an 8-bit grayscale image when drawing the Moon and deep sky objects or a 24-bit RGB color when drawing noticeably colored objects like Mars and Jupiter, colored double stars, aurorae, and so on. When completed, the images can be saved as JPEG files with no compression, in order to preserve as much image quality as possible. An 8-bit grayscale 300 dpi A6-sized drawing of the Moon may be around 1 MB in size, while a 24-bit RGB image of the same subject might amount to twice the size.

Graphics Programs for Cybersketching

Name	Publisher	Platform
GIMP	**GIMP**	**Win/Mac/Linux (freeware)**

http://www.gimp.org

GIMP (GNU image manipulation program) is a free application with a great deal of functionality. It's well worth downloading even if you already have a full-featured commercial graphics program. GIMP is suitable for a variety of image manipulation tasks, including photo retouching, image composition, and image construction. It can be used as a simple paint program, a high-quality image retouching program, an image format converter, and so on. The program is expandable, designed to be augmented with plug-ins and extensions to perform just about any graphics tasks (Figure 7.9).

HandPainter	**SoftAway**	**Palm**

http://www.softaway.com

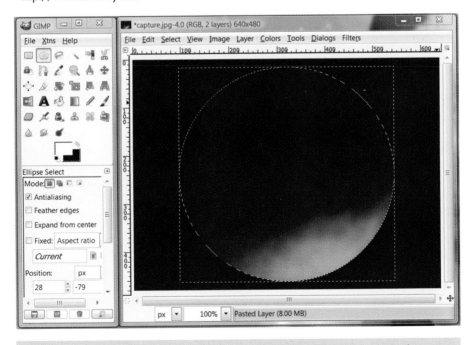

Figure 7.9. Screenshot from *GIMP* showing an enhancement of a lunar eclipse observation (photo by Peter Grego).

HandPainter 0.6 is a relatively simple freeware graphics program for Palm OS featuring the basic functions required to produce a quick astronomical cybersketch. Its big brother, *HandPainter Pro*, is a more sophisticated commercial program worth taking a look at (Figure 7.10).

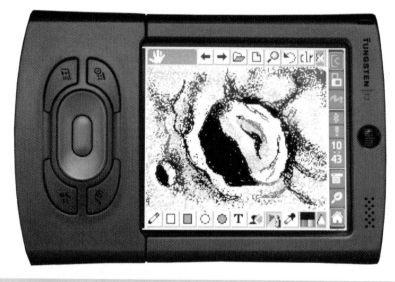

Figure 7.10. A straightforward cybersketch of a lunar crater, made on a Tungsten T3 with the relatively simple Palm graphics program *HandPainter 0.6* by SoftAway (photo by Peter Grego).

Mobile Atelier/Mobile Pencil	NeFa	PPC

http://www.freewareppc.com/graphics/mobileatelier.shtml

A freeware cybersketching application with an easy-to-use, plug-in-based interface. All functions are provided as a 'plug-in' format, and the user can select which functions to use. *Mobile Atelier* is not the most highly sophisticated graphics program in the world, but its various drawing functions, including variable-sized and shaped brush strokes, blur, smudge and contrast controls, plus the ability to work in layers, make it suitable for quick, no-fuss cybersketches. Great for lunar and planetary sketches. A simpler freeware program by NeFa, *Mobile Pencil*, features eight types of pencil tip and different types of paper texture and is a delight to use (Figure 7.11).

Photopaint	Corel	Win/Mac

http://www.corel.com

Photopaint is a full-featured raster graphics editor bundled with the *CorelDRAW* graphics suite. The current version, *Photopaint X4*, is a versatile program

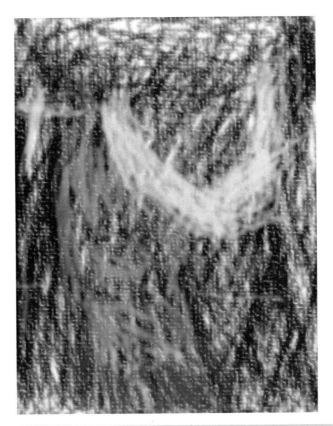

Figure 7.11. 'Rocket to the Moon' – an imaginative cybersketch made on *Mobile Pencil* by Jacy Grego, the author's daughter (then aged $3^3/_4$ years) (photo by Peter Grego).

particularly suited to cybersketching with graphics tablets and tablet PCs. *Photo-paint* (version 8) is one of the simplest and most intuitive interfaces. *CorelDRAW* itself also provides a similar ease of operation, and for cybersketching-related purposes it is particularly suitable for preparing observing blanks, reports, and exhibition material (Figure 7.12).

Photoshop	Adobe	Win/Mac

http://www.adobe.com/products/photoshop

Photoshop has been enjoyed by Mac users since the first version was released in 1990; the first Windows version appeared a couple of years later, and since then it has gone through eight more versions, becoming one of the world's leading graphics programs in the process. The program was renamed *Adobe Photoshop CS* in 2004 – an abbreviation shared by other programs in Adobe's Creative Suite range. Files created in *Photoshop's* native format (with the *.PSD extension) can be

Figure 7.12. A poster advertising the BAA Lunar Section, prepared by the author in *CorelDRAW* for display at the BAA Exhibition Meeting in 2008 (photo by Peter Grego).

readily imported into all the other Creative Suite programs, including *Adobe Illustrator CS, Adobe ImageReady CS,* and *Adobe Acrobat. Photoshop* can be used to edit numerous *raster* and *vector* image formats.

PhotoPlus X2/DrawPlus X2	Serif	Win

http://www.serif.com

Serif claims to be the world's leading independent publisher of desktop publishing, design, and graphics software. Two of its flagship graphics programs, *PhotoPlus X2* and *DrawPlus X2*, are of considerable interest to the digital imager and cybersketcher. At one point Serif offered free downloads of full versions of graphics software, including *PhotoPlus 6* and *DrawPlus 4* (among other free software); these could be obtained from http://www.freeserifsoftware.com, so the user could decide if the programs were suited to their imaging and cybersketching requirements. You can check it out and see if this offer is still available.

Paint Shop Pro/Painter X	Corel	Win

http://www.corel.com

A powerful photo editor and graphic design tool, *Paint Shop Pro* was originally developed by Jasc Software. Version 8 of the program is a very capable program, particularly for desktop work. In 2004 Corel Corporation – already a giant in the computer graphics industry – took over Jasc and continued to develop several lines of its products, including *Paint Shop Pro*. However, the latest version of this popular software, *Paint Shop Pro Photo X2*, is aimed squarely at the photographer and offers a wide range of photo editing tools. Its stable mate, *Painter X*, now takes the place of *Paint Shop Pro* as Corel's premier art software. It's claimed that *Painter X* simulates traditional painting better than any other software through the use of unique digital brushes, art materials, and textures in a complete digital art studio.

PDAcraft Paint	PDAcraft	PPC

http://www.pdacraft.com

One of the more basic cybersketching programs, with a simple and pleasant interface and the ability to produce drawings of up to 1000 × 1000 pixels (Figure 7.13).

Photogenics	Idruna	PPC

http://www.idruna.com/products_pocketpc.html

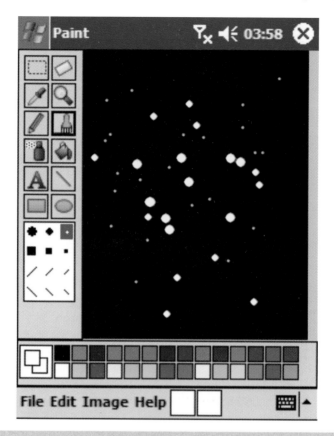

Figure 7.13. Observational cybersketch of the central region of the open star cluster M44, drawn with PDAcraft Paint on PDA (photo by Peter Grego).

Photogenics is a very good, full-featured graphics editing program that offers the ability to produce cybersketches in a variety of realistic media, such as airbrush, chalk, pencil, paintbrush, watercolor, and sponge. More than 60 filters of various sorts – many of which are useful to the astro-imager and cybersketcher – can be applied either globally or to selected areas of the image to provide instant enhancements. Paint layers enable mistakes to be erased with a 'fade out' technique without having to modify the best parts of an image. The program also offers a wide range of special effects, including paint-on pyrotechnics; although most of these are not really relevant to astronomical cybersketching, small flare effects can be used to reproduce stellar 'bursts' to add a little aesthetic interest to starfield images (Figure 7.14).

Pocket Artist	**Conduits**	**HPC/PPC**

http://www.conduits.com/products/artist/

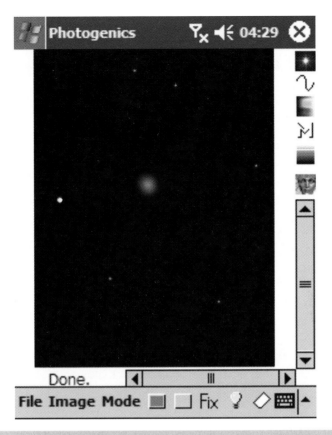

Figure 7.14. Observational cybersketch of the galaxy M49, drawn with *Photogenics* on PDA (photo by Peter Grego).

One of the fullest-featured and most versatile of mobile graphics editors, *Pocket Artist 3* is a 24-bit color graphics program that can be run on PDAs and handheld devices operating under Windows Handheld PC 2000, Pocket PC 2000, Pocket PC 2002, and Windows Mobile 2003. It provides a comprehensive set of image editing tools, including a range of painting tools such as paintbrush and pencil (with sketch mode), paintbucket and quick shape tools, text input, color, and brightness and contrast adjustment along with noise and unsharp mask filters. Files in PSD (Adobe *Photoshop's* native format), JPEG, PNG, GIF, BMP, TGA, and J2K format can be opened, edited, and saved (Figure 7.15).

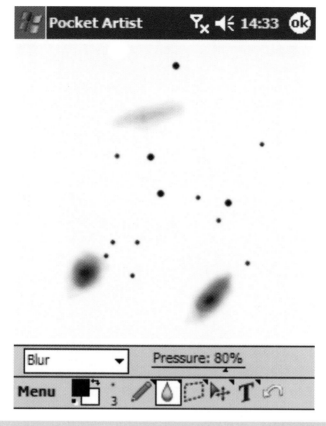

Figure 7.15. Screen capture of an observational drawing made in Pocket Artist, showing Leo galaxy triplet M65, M66, and NGC 3628 (photo by Peter Grego).

Here, now, are some other graphics programs of note:

Name	Publisher	Platform
Imageer	SPB	PPC
Jinzo Paint	t-ueno	PPC (freeware)
Pocket Paint	Microsoft	PPC/HPC (freeware)
Pocket Painter	Aidem	PPC
Sketch	Lingle Tech	Palm
Sketcher	Palmetto	Palm (freeware)
VSpainter	Virtual Spaghetti	PPC
UltraG	Masataka	PPC (freeware)

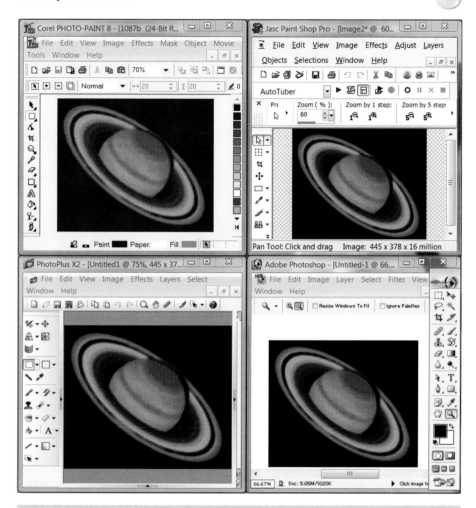

Figure 7.16. Screenshot of four different graphics programs open concurrently and showing the same image – *PhotoPaint 8, Paint Shop Pro 8, PhotoPlus X2* and *PhotoShop 7* (photo by Peter Grego).

Screen Capture

It's often very useful to capture the image displayed on part or all of the computer screen (be it a PC monitor or PDA screen) and save it as a graphic image (Figure 7.16). Indeed, many of the images in this book have been obtained using screen capture programs. For cybersketching, some kinds of observing templates can be prepared fairly quickly and easily by capturing a screenshot from an astronomical program. For example, a screenshot from a program displaying an accurate rendition of Saturn and its tilted rings can be imported into a graphics program, cropped, and modified (depending on the original image) to be used as an observational template – all while in the field at the telescope eyepiece.

Some screen capture utilities are simple and unsophisticated, creating a medium-resolution bitmap of the whole screen and copying it to the clipboard to be pasted into any graphics application of choice. Multimedia keyboards have a special 'Prt Scr' (screen capture) key that does (or is supposed to do) exactly this, while pressing 'Alt' and 'Prt Scr' simultaneously captures the active window. Other screen-capture programs permit the user to define hot keys to activate the capture, to select specific areas on-screen to capture (including shaped areas), and to determine the file format and resolution of the captured image.

Here are some screen capture programs:

Name	Publisher	Platform
Capture Me	Chimoosoft	Mac (freeware)
CECapture	Epiphan	PPC/HPC (freeware)
Magic SS	Terrailloune	PPC (freeware)
MWSnap	Mirek	Win (freeware)
PDAcraft Capture	PDAcraft	PPC (freeware)
PocketScreen	Rhapsody	PPC (freeware)
ScreenHunter	Wisdom	Win (freeware)
ScreenShot	SPB	PPC (freeware)
Snap	Maillot	Palm (freeware)
vSnap	Wong	PPC (freeware)

Basic Drawing Tools

Most graphics programs share a common set of basic drawing tools, although not all have similar-looking icons. For simplicity sake let's use the basic toolbox in *PhotoPaint 8* as our guide, to discover what they do and how they can be applied to astronomical cybersketching (Figure 7.17).

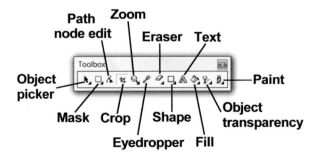

Figure 7.17. Basic toolbox in *PhotoPaint 8* (photo by Peter Grego).

Paint Tool

Represented by a brush icon, the paint tool is the most important tool in the box. Of the four functions accessible by clicking on this button, chief among them is the paintbrush, for which a wide range of artistic styles can be selected, including custom pen (ballpoint, finepoint, etc.), traditional paintbrush (oil paint, watercolor, etc.), airbrush, and power sprayer (Figures 7.18 and 7.19). All of these artistic styles are fully adjustable in terms of their size, shape, texture, intensity, transparency (the amount of 'see through'), and they can also be modified in terms of the way in which the chosen effect interacts with the background. Brushes can be customized to any desired parameters and saved for later use.

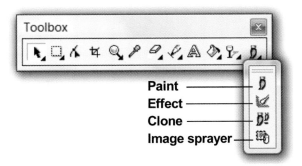

Figure 7.18. The paint tool's main functions in *PhotoPaint 8* (photo by Peter Grego).

In addition to the paint tool, three other functions are accessible by clicking on the paint button – effect, clone, and image. Effect is a very important tool, because it enables the cybersketcher to make adjustments to selected parts of the image in a number of interesting and useful ways, including smearing, smudging, brightness, contrast and hue adjustments, sponging, tinting, blending, sharpening, undithering, and dodging/burning. You may, for example, want to use quite a bit of smudge, smear, and undithering on your observational lunar drawings when enhancing the cybersketch on the desktop PC to achieve the desired results, which produces a characteristic 'look' to the images. However, the extent to which the user might choose to apply each of these effects is, of course, a matter of personal taste. You should experiment as much as possible with your desktop and portable computer's graphics programs, using relatively simple graphics, to discover how each effect can be used and finely controlled.

The clone tool (sometimes termed a rubber stamp) is used like a paintbrush and enables a selected area of the cybersketch to be replaced with a portion of image data from another part of the sketch. Cloning is really useful for retouching images, invisible repairing, and for filling in small areas of a cybersketch, say,

Pen   Brush   Airbrush   Sprayer

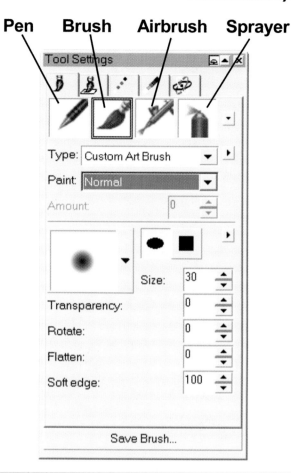

Figure 7.19. The four main paint effects in *PhotoPaint 8* (photo by Peter Grego).

where there's a blank portion of the sketch that was neglected in the original field drawing (Figure 7.22). The other tool mentioned above, the image sprayer, provides a set of interesting graphic functions, but it's actually of no practical use for general astronomical cybersketching.

In addition to being able to change the type of effect and the size and shape of the brush tool, its intensity, transparency, edge softness, and anti-aliasing can be adjusted to the user's preferences. Transparency is useful for producing brush strokes with various degrees of 'see through,' and it's a useful attribute for brushwork where it may be necessary to add tone to

Regular (solid paint)

Faded edge brush

Dry camel hair

Soft wet oil

Calligraphy pen

Spray can

Watercolour

Soft chalk

Pontillism

Impressionism

Airbrush

Figure 7.20. Graphics programs offer a variety of brush options. Here are just a few, from *PhotoPaint 8* (photo by Peter Grego).

areas containing fine detail that would otherwise be lost or changed using an opaque brush – notably in lunar and planetary cybersketching. Modifications to the sharpness of brushstroke edges can be done by changing the edge softness attributes and applying anti-aliasing. This is another area where the user ought to experiment with graphics software to discover its capabilities. There are many routes to the same destination in cybersketching.

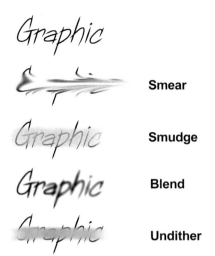

Smear

Smudge

Blend

Undither

Figure 7.21. Showing different effects applied to the same graphic (photo by Peter Grego).

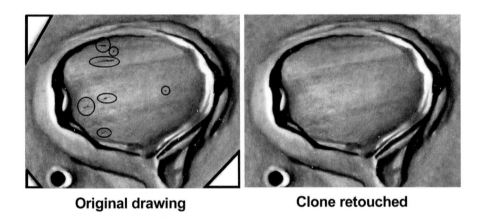

Original drawing   Clone retouched

Figure 7.22. This regular pencil sketch of Archimedes crater, scanned from the original drawing, shows where it is unfinished in three corners and displays numerous pencil scuff marks. Cloning has filled in the missing three corners of the drawing and has removed the unwanted artifacts to produce a much nicer enhanced image (photo by Peter Grego).

Filters

Some of the effects mentioned above can be applied globally or to selected parts of an image to significantly enhance cybersketches with minimal effort. Among the most-used filters in cybersketching are ones that control image brightness and contrast; these filters remove image artifacts and adjust the image sharpness.

The brightness and contrast levels within a cybersketch are, up to a certain point, a matter of personal preference. However, it's worth pointing out that what is seen on one screen is not always what you (or others) see on another monitor, nor might it correspond with the printed image. Frequently the brightness, contrast, and gamma level settings on-screen are such that the highest range of brightness or the lowest range of darkness cannot be differentiated sufficiently when making adjustments within these ranges. This can lead to some embarrassing results in which some aspects of the image editing process, such as erasing, cropping, and pasting, brightening, and darkening brush artifacts and the like are clearly visible (Figure 7.23).

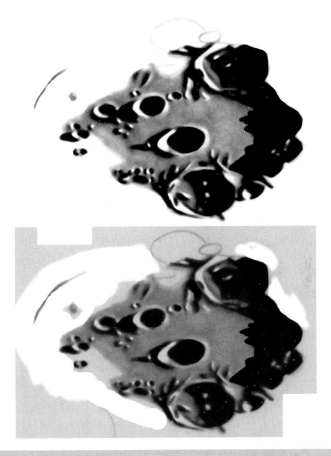

Figure 7.23. This cybersketch of the crater Clavius shows how image editing is not always invisible in the end result. At top, the edited image is shown satisfactorily on a poorly adjusted monitor whose high and low grayscale values, in terms of brightness, contrast, and gamma, are poorly differentiated. However, a properly corrected monitor shows up numerous editing artifacts on the very same image (photo by Peter Grego).

A very useful tool called tone curve allows fine, freeform adjustments to be made over the range of an image's grayscale or color tones. The range of possible intensities, from darkest to lightest, is displayed to the side of and across the bottom of the graph. A diagonal line across the tone graph, from lower left to upper right, represents an unchanged image (Figure 7.24). Tone curve can enhance an image beyond measure. Careful manipulation of the curve can darken the darkest pixels and brighten the brightest areas without affecting the intensity of the image's mid-tones; similarly, the mid-tones can be adjusted without affecting the extremes. The effect can remove some types of image editing artifacts by shifting the image's tonal range to the desired levels.

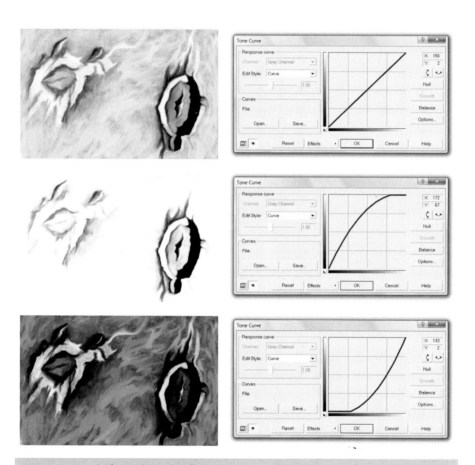

Figure 7.24. The figure shows the effects of tone curve adjustments on a cybersketch of the lunar crater Reiner and the swirl Reiner Gamma (photo by Peter Grego).

Blurring and Sharpening

No discussion of the basics of image enhancement is complete without mentioning blurring and sharpening filters. The most frequently used blur effect, called Gaussian blur, reduces an image's noise and detail levels, producing the effect of viewing the original image through a sort of translucent screen. Among its various astronomical cybersketching uses, Gaussian blur can be successfully applied to drawings of nebulous deep sky objects and depictions of vague albedo features on roughly sketched lunar landscapes (Figure 7.25).

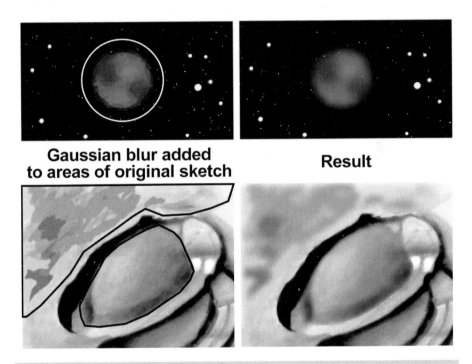

Gaussian blur added to areas of original sketch

Result

Figure 7.25. Gaussian blur, applied to observational cybersketches of the Dumbbell Nebula and parts of the landscape in and around the crater Steinheil (photo by Peter Grego).

Gaussian blur can also be used to remarkably good effect on halftone or low-resolution monochrome images in order to produce stunning transformations of images that may have once been considered beyond redemption. For example, it can be used to significantly enhance and improve observational drawings of Mars made during the 1988 apparition using the crude spraycan drawing software on an Apple Macintosh (Figure 7.26).

Another sort of blurring effect, the median filter, works by removing noise but retaining some of the detail (confusingly, *GIMP's* median filter is called

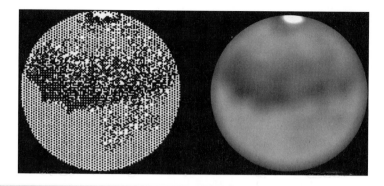

Figure 7.26. Gaussian blur transforms this vintage observational cybersketch of Mars, made on an Apple Macintosh in 1988, into a pleasing cybersketch (photo by Peter Grego).

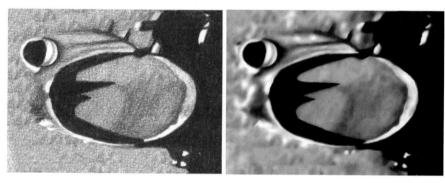

Scanned pencil sketch   **Median filter applied, plus tone curve adjustment**

Figure 7.27. Filters can be applied to scanned observational drawings. Here, a median filter and tone curve adjustment has been made on a scanned pencil sketch of the lunar crater Plato, producing a much nicer image without sacrificing any important detail (photo by Peter Grego).

Despeckle). Median is very useful in cybersketching, but if applied too heavily it can produce unwanted artifacts of its own that can require removing. Median can be applied to smooth out the appearance of a cybersketch, and when used on scanned pencil sketches it can remove unwanted pencil strokes, graphite grains, and coarse paper textures to create a smoother look (Figure 7.27). Median is a useful filter to apply when enhancing low-resolution PDA sketches of star fields, where the individual stellar points display unwanted artifacts. Median smoothes out the edges of each star, producing more coherent rounded dots (Figure 7.28).

Some aspects of cybersketching call for the use of scans or photographic images as drawing templates. Depending on the source and quality of the

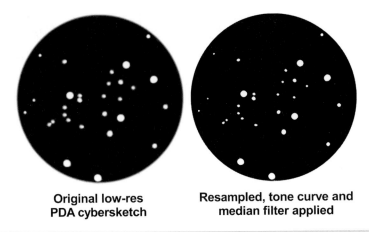

**Original low-res
PDA cybersketch**

**Resampled, tone curve and
median filter applied**

Figure 7.28. A low-resolution observational PDA cybersketch of the star cluster NGC 6709 is enhanced at the desktop by resampling to a higher resolution, applying a steep tone curve to remove grays and finally using a median filter (photo by Peter Grego).

image used, it may be necessary to enhance the image so that it is more suitable for use; this is where filters to remove speckle, dust, scratch, and moiré patterns come in very handy. Speckle, noise, and 'snow' artifacts are caused by unwanted pixels; these artifacts, plus distracting particles of dust and scratches on photographic images, can be automatically removed without excessive degradation of the image quality. Unsightly moiré patterns, produced by interference effects between lines in an image or halftone pattern, are often seen in scans of printed images in books and magazines (Figure 7.29). Moiré patterns can be removed

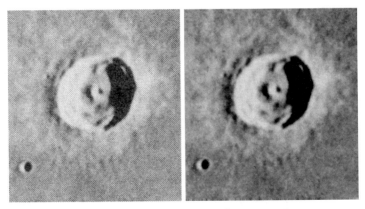

Before and after Moiré removal

Figure 7.29. Scans of printed halftone images such as this can display unwanted moiré patterns; these can be effectively removed with graphics software (photo by Peter Grego).

very successfully, and the anti-moiré tool can also help to remove or reduce the graininess of scanned pencil drawings.

Going to the other extreme, now, a process known as unsharp masking – whose origins lie in the traditional photographic darkroom – can be used to sharpen an image in an almost magical way. Of course, there's no hocus-pocus here. The process works by creating a duplicate image of the original, blurring it, and adding the difference between the two to create a sharper result. What might be considered poor, blurry, and indistinct astrophotographic images can be enhanced with unsharp masking to produce pleasant-looking images nearer to reality (Figure 7.30). Unsharp masking isn't only useful when applied to astrophotographs. Cybersketches can also benefit from its application in a number of ways (Figure 7.31).

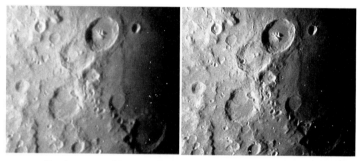

Unsharp masking applied to a CCD image of the Moon

Figure 7.30. Unsharp masking, applied to a CCD image of the lunar craters Theophilus, Cyrillus, and Catharina (photo by Peter Grego).

PDA cybersketch, before and after unsharp masking

Figure 7.31. Unsharp masking has been applied to this observational cybersketch of the globular cluster M15, made on a PDA. All the details are preserved, and individual stars are more clearly defined in the resulting image (photo by Peter Grego).

Fills

Traditional pencil sketching requires quite a bit of work to fill large areas of a drawing with an even tone. Extensive black areas (the lunar shadows, for example) are particularly troublesome for the conventional sketcher and require patient and even application of several layers of graphite to produce a dark (but never black) tone. Areas of lighter tone may display individual pencil strokes, whose removal might be accomplished by the use of finger smudging – it's a bit of a messy business. No such problem exists in the realm of cybersketching. Expanses of single tone are easily created by defining an area with the 'mask' tool and applying an appropriate filter, say the adjustment of tone levels. Alternatively, the 'fill' tool may be used to convert all the pixels within a defined area to an identical tone. Various fill effects are also easily accomplished, such as gradient fills that are particularly useful for creating near-terminator shadow effects in a lunar sketch.

Working in Layers

Layer work entails creating different levels within an image, each of which can be given different attributes and which can be worked upon and modified without necessarily affecting the information contained in the other layers. Although it is capable of producing pleasing results when applied to astronomical cybersketching, layer work does require some time to discover its capabilities and to become proficient in its use. Layer work is usually stored in the graphics program's native format and can be transformed into a single combined image when saved in a regular bitmap format, such as JPEG.

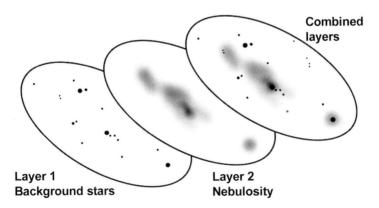

Figure 7.32. Using a layer to add nebulosity to a PDA cybersketch (photo by Peter Grego).

Layer work can be used to produce cybersketching templates from existing digital images, using it as a form of sophisticated tracing paper to produce accurately delineated observations; this is particularly useful in lunar observation where there is a copious amount of detail that must be sketched in the right proportions. Layer work is also very useful for adding substance to a background image without affecting its detail. For example, wisps of nebulosity can be added to a detailed cybersketch of a deep sky observation without blurring the background stars (Figure 7.32).

Resampling and Resizing

As we've already seen, images prepared in many PDA graphics programs aren't of a spectacularly high resolution, although some programs do enable the user to select a specific image size in pixels (within limits), allowing finer work to be achieved on the small screen. By default, the freeware P/PC program *Mobile Atelier* saves images in a BMP format at 72 dpi with dimensions of 85 × 113 mm.

Some observers will be completely content with saving the original low-resolution cybersketch as their only record of the observation – and there's nothing at all wrong with this. Still, when it comes to transferring PDA sketches to the desktop PC and viewing them on a large monitor, the images may look awfully crude and blocky when compared to that delicately detailed little depiction you created on the PDA. So, if you do intend to go further and enhance your observational cybersketches, it is always best to get to work on them at the desktop computer as soon as possible after the observing session, while the memory of the observation remains fresh in your mind.

After you've made your drawing on the PDA at the telescope eyepiece, remember to retain its orientation when you're making those finishing touches indoors at the desktop computer. It's best not to immediately flip the cybersketch north–south or east–west or rotate the image to conform to the normal convention before commencing work. Your mind's eye will have spent maybe an hour or more absorbing the image as it presented itself at the eyepiece, and many of the details of a drawing remain in the mind's eye for some time after the sketch is made; a flipped or rotated image is likely to cause your mind's eye a little confusion. While working on enhancing the cybersketch in the orientation in which you observed it, you will more easily remember subtle details and the 'lie of the land,' bringing to mind many of the visual clues depicted in the sketch. Naturally, it's the scene that you observed and depicted that is nearer the truth, so don't try to second guess your own psychology!

To enhance low-resolution PDA cybersketches it is necessary to resample them to a greater resolution – 300 dpi being the preferred standard. This initial step will remove a little of the image's coarseness, but further work may need to be performed to produce a cybersketch you're happy with; this is where the enhancements and filters mentioned above come into play.

It will probably be necessary to crop your cybersketch to the desired size by using a rectangular selection tool; this can be adjusted using small 'handles' visible at the edges of the selection box. Images can also be flipped top to bottom or left to right in order to conform to a particular orientation, say if a star diagonal was used to make the observation and you want the end result to appear in its true orientation. Images can also be rotated to any desired angle to produce the desired result. To avoid producing jagged edges in the rotated image it's advisable to use anti-aliasing when doing so.

End Game

The enhanced cybersketch, of which you're justifiably proud, can be used in a number of interesting ways (limited only by your imagination). Here are a few examples:

- Your cybersketch may be put on a web site, reduced to a lower-resolution 'thumbnail' version on the web page. You may decide to make the full image itself available for people to view and download by hyperlinking it to the thumbnail image.
- Information about the observation and the objects it portrays can be added in text, around the cybersketch itself and/or made directly onto an adjacent copy of the cybersketch.
- To make an interesting comparison, the finished cybersketch may be shown either adjacent to the original sketch made at the eyepiece, next to the digital image template used in its preparation (if a template was used), or beside a computer-simulated graphic of the object in question.
- The completed cybersketch might be incorporated electronically into a standard observing form used by any of the large amateur astronomical organizations (such as ALPO in the United States, the BAA and SPA in the United Kingdom). This can be done by importing the image directly into a PDF format of the observing form or pasting it into a high-resolution scanned bitmap of the form.
- Cybersketches can be arranged into image albums, viewable on the computer monitor in a number of ways. They can be grouped by feature type, date, observation, etc. There are many freeware and commercial programs capable of arranging your electronic images into albums.
- Display your cybersketches in a digital photo frame.

The possibilities are endless.

Vector Graphics

Vector graphics can be used to draw circles, ellipses, and geometric shapes of one kind or another, so they are very useful for preparing astronomical observing blanks and observing forms. Because vector images are independent of resolution, they can be scaled without any loss of image quality. Planets come in a variety of shapes, so vector graphics are ideal for outlining them in a precise manner, based upon ephemeris data giving the planets' phases and tilts. Examples of special shapes that can be constructed using vector graphics include the oblateness of the Jovian disk, Saturn's oblate disk and the elliptical outlines of its ring system (which vary according to the planet's tilt toward us), the varying gibbous phase of Mars, and the more extensive phases of the inner planets Mercury and Venus (Figure 7.33).

Figure 7.33. A variety of planetary shapes created using vector graphics.

CHAPTER EIGHT

Cybersketching Challenges

Now let us delve into more specific detail concerning the various techniques used to make cybersketches or to digitally enhance astronomical observational drawings of one kind or another.

Observational Field Drawings

Sketches have been made at the telescope eyepiece ever since Thomas Harriott first peered at the lunar surface through his little refractor back in the summer of 1609 and attempted to depict what he could see of that hitherto mysterious orb. Most of the great astronomers of the seventeenth, eighteenth, and nineteenth centuries tried their hand at observational drawing, with varying degrees of success.

The technology now exists to significantly enhance the powers of the visual observer (not to mention increasing the enjoyment factor) by using computers at the eyepiece – a development that would likely have been welcomed by the likes of Galileo, William Herschel, and Eugene Antoniadi (Figure 8.1). Cybersketching is a natural development of traditional observing techniques.

Actually, there is an interesting century-old corollary with cybersketching. Back in the late nineteenth century, amateur astronomer Johann Krieger (1865–1902) used a substantial inheritance to build his own observatory equipped with a 257 mm (10.1 in.) refractor in the suburbs of Munich, the Bavarian capital. His chief astronomical pursuit was lunar observation, and possessing no mean degree

P. Grego, *Astronomical Cybersketching*, Patrick Moore's Practical Astronomy Series, DOI 10.1007/978-0-387-85351-2_8, © Springer Science+Business Media, LLC 2009

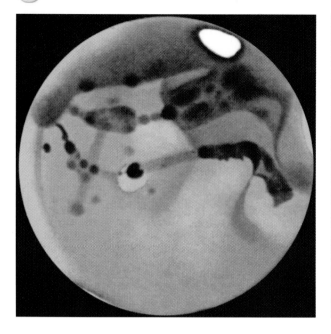

Figure 8.1. Eugene Antoniadi, one of the greatest observers of the late nineteenth and early twentieth centuries, was renowned for his superb renditions of Mars. This observational sketch was made around the date of the planet's 1909 opposition.

of artistic talent, he set his sights on producing a detailed map of the Moon. To this end, Krieger obtained a set of photographs taken through the 91 cm (36 in.) Lick and 83 cm (33 in.) Paris Observatory refractors. Enlargements of sections of these photographs, printed at low contrast, were used by Krieger as observing blanks for his visual observations. Over the course of several observing sessions, detail was added in various media, including graphite pencil, charcoal, and ink to produce accurate and incredibly beautiful depictions of the Moon that stood out from most other visual studies of the era. The first 28 of these exquisite lunar studies were published in Krieger's *Mondatlas* (1898), with further work of his being published posthumously during the twentieth century (Figure 8.2).

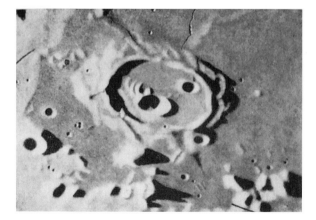

Figure 8.2. Johann Krieger's depiction of the lunar crater Cassini, from *Mondatlas.*

Sadly, few other observers chose to follow up on Krieger's promising techniques, preferring instead to continue depicting features as well as their hand–eye–brain coordination would allow. Admittedly, many amateur astronomers are very good at this, producing work whose positional accuracy isn't that far off, but some find it difficult to translate the view through the eyepiece into an accurate sketch. It's not simply a question of artistic talent – after all, it's evident that many famous modern artists have absolutely no ability to draw accurately – but rather more to do with technique.

Observational drawings demand careful attention to the way in which the sketch is laid out from the outset. The biggest mistakes are made early on in a drawing, where adding detail to an inaccurate base simply compounds the problem. There are bound to be problems if a sketch is made by working from one corner and hoping that everything looks right when the other corner of the page is eventually reached.

Patience is a virtue of great importance in observational astronomy. When observing the Moon, the observer must decide on the area within the eyepiece's field of view that will form the focus of their drawing, setting definite boundaries to the sketch. Deep sky observers often attempt to depict everything in the eyepiece's field of view, with the deep sky object itself centered in the field, while planetary observers will of course want to depict the entire planet in one session (or, taking a broader view, depict the configuration of its satellites).

Working to a realistic scale is important. Users of PDAs will have no problem in filling up the entire screen with a cybersketch – after all, it measures just a few tens of square centimeters in total, a size that is perhaps a little smaller than one might choose to make an actual observational sketch with pencil and paper. However, the zoom tool can be used to magnify and work upon any areas wherever it's necessary to add detail too fine to accurately portray in the regular display mode. It goes without saying that it's important to ensure that the touchscreen on mobile devices ought to be calibrated as accurately as possible (see above); otherwise, you may find yourself using the zoom tool a little more than really necessary in order to rectify mistakes.

Because UMPCs and tablet PCs have more generously sized screens than PDAs, it may be tempting when using these devices to do most of the work on a field sketch that is zoomed-in so that its edges fill the entire screen. However, a much more comfortable and efficient size for laying out a cybersketch is within the area defined by the $90°$ arc, which can be drawn by a single stroke of the stylus without moving the arm, the end points of the arc determining the diagonal proportions of the sketch. This is about 11 cm on the diagonal, or around 7×9 cm (landscape), which means that it fits nicely on an A6-size page (10.5×14.8 cm/4.1×5.8 in.) set to 300 dpi with a little room to spare around the margins. Finer details on the cybersketch can, of course, be added by zooming in to specific areas.

Observational cybersketching covers a variety of techniques. If you're used to making conventional pencil sketches, some of these techniques may have a more familiar and user-friendly feel, at least at first, than others.

PDA Cybersketching

Despite their small screen size, PDAs offer much to the visual observer. They're eminently portable, comfortable to hold in the hand for long periods of time, and have a pretty long battery life. We've seen how astronomical software can be used in the field to plan observations and produce observing blanks and how graphics software is capable of translating stylus strokes on their touchscreens into sound astronomical observational sketches. Button mapping is a useful tool in which individual applications can be assigned to any of the PDA's physical buttons, allowing programs to be accessed quickly and viewed at the user's convenience. For example, buttons on the PDA may be assigned to begin screen capture, planetarium, and drawing programs.

Lunar cybersketches made on a PDA may well take up the entire screen area, or a substantial portion of the central area of the screen. Planetary observations are best made using a pre-prepared blank showing the planet's basic outline. The standard traditional sketching template size for observations of Mercury, Venus, and Mars is 50 mm (2 in.) in diameter. Jupiter is outlined by an ellipse with a major axis of 63 mm (2.5 in.) and a minor axis of 59 mm (2.3 in.). Saturn's oblate sphere is represented by a major axis dimension of 44 mm with a ring span of 102 mm (the outer edge of ring A). All of these will fit nicely within any PDA screen.

Cybersketches of star fields are best made on a circular template – a negative image with a white background and black stars works best – with its diameter set almost to the width of the PDA screen and its upper edge near the top edge of the screen. This will allow written notes to be made below the drawing, either freehand notes or actual text input onto the screen.

On its regular setting, the PDA's backlit screen will probably prove too bright to be comfortably used in the field for anything but solar and lunar work. However, the screen brightness on PDAs is adjustable, usually to tolerably low levels so that the eye's dark adaptation can be maintained. Sometimes, though, a little extra dimming of the screen image is needed, and one method is to make the cybersketch on a template with a deep red background rather than a white one, sketching the stars as black dots and nebulous areas as shades of gray in a negative image. You can also physically mask the entire screen area with a sheet of thin red acrylic material or acetate so that changes between programs don't ruin the eyes' dark adaptation.

If the PDA has a voice recording facility (and most of them do), cybersketches can be accompanied by spoken notes, adding quite a lot to an observation's scientific worth. Voice notes are useful for calling out any features that aren't able to be sketched accurately or for pointing out any interesting or unusual aspects of the observation. Voice notes also help when copying-up and enhancing observational cybersketches on the desktop computer, increasing their accuracy the longer the period separating the original observation and the enhanced drawing.

Making Freehand Field Cybersketches of Lunar Features on a PDA

A wealth of lunar detail is visible through the telescope eyepiece. Within the observer's visual grasp are many thousands of craters, ranging from ancient lava-flooded impact basins to relatively young impact craters with bright ray systems. Mountains, hills, domes, rilles, clefts, faults, and valleys all jostle for space in the crowded lunar landscape. Lunar observation is by far the most visually rewarding branch of astronomy. The constantly changing vistas of the Moon's surface are every bit as visually stimulating as the contemplation of an impressionist painting. The beauty is that the Moon belongs to everyone and doesn't cost a penny to look at.

Most lunar observers regard the telescope eyepiece as if it were the porthole of their very own Apollo command module. The privilege of just seeing is satisfying enough, yet ever since Galileo sketched the lunar craters nearly 400 years ago, many observers have striven to keep some kind of permanent record of their forays around the Moon's surface.

Those who take the opportunity to make lunar drawings will discover an activity that improves every single aspect of their observing skills. The Moon is packed with very fine detail, and the ability to discern this constantly improves with hours spent at the eyepiece. During a course of lunar 'apprenticeship' the apparent confusion of the Moon's landscape becomes increasingly familiar.

Beginners to lunar observation are best advised to begin finding their way around the Moon by attempting to draw features that lie near to the terminator, the slowly shifting dividing line between the Moon's daytime and nighttime sides. Here, lots of shadow is thrown out by the Sun's low elevation, and plenty of detail can be seen.

Do try to have confidence in your own drawing abilities. The lunar observer isn't some kind of weird nocturnal art student, and marks aren't given for artistic flair or aesthetic appeal. Observational honesty and accuracy counts above all. To improve your lunar cybersketching skills, practice by drawing sections of lunar photographs that appear in books and magazines.

When the Moon is sharply focused in the eyepiece, don't be intimidated by the sheer wealth of detail. Find your bearings with a good map of the Moon. Only small areas should be chosen to sketch, say an individual crater, and at least an hour should be devoted to the study. It's best to identify the feature being observed while at the eyepiece. If it can't be immediately identified, note its position relative to larger features that can be identified.

Here's an example of a quick, freehand lunar field sketch made without a template. It's typical of the sort of lunar observation that might be made without too much effort, where it might be impractical to use a tablet PC, say when observing from a remote site on a camping trip, or observing through someone else's telescope. In this particular case it was a cold Christmas night following a family get-together. The Moon provided a source of visual entertainment more enthralling than any amount of television and more mentally stimulating than playing games.

Date: December 26, 2007
Time: 00:55 – 01:35 UT
Feature: Fraunhofer
Instrument: 127 mm MCT (Meade ETX-125) ×200
Cybersketching device: SPV M2000 P/PC
Cybersketching programs: *Mobile Atelier* (field sketch) and *Corel PhotoPaint*
(desktop enhancement using a graphics pad and mouse)

Fraunhofer, a 57 km diameter crater, lies near the Moon's southeastern limb. With the Moon a waning gibbous phase aged 16.3 days, the feature's illumination came from an evening Sun. Fraunhofer's western rim was casting an internal shadow but most of the details on the crater's floor remained visible, making a perfect subject for an observational drawing.

A small area including Fraunhofer and the craters Fraunhofer A and H was selected for the cybersketch – not too broad an area, presented fairly generously at a magnification of 200×, and one that was thought possible to depict sufficiently well within the space of 45 min.

1. *Mobile Atelier* was used to make the cybersketch – a good program for quick, fuss-free depictions. A portrait format and smooth white background was decided upon. Initial depictions of the outlines of the main topographic features were made using a narrow brush with a light gray color, beginning with Fraunhofer itself and then adding the features in and around it. Small mistakes in delineation were not considered important at this stage, as they could be overdrawn later. The drawing was saved on the PDA's SD memory card.

2. Still using a narrow brush, the darker areas of shadow were outlined and filled in. This could also be achieved using a fill tool for more consistency of tone, but since the black areas were fairly small and the drawing was made on a PDA, it was decided to stick with brush strokes. Switching color to a light gray, the general tone of the landscape was depicted, leaving brighter areas such as Fraunhofer's inner eastern wall and outer western flanks unshaded. Some albedo detail was added, such as the dusky banding on Fraunhofer's inner wall. The drawing was saved.

3. Further tonal detail and shading was added, with some variations in the brush tone, and the drawing was saved.

4. Happy that the main features were adequately depicted, a blur effect was applied to the brush, and a smoother consistency of tone in and around Fraunhofer was achieved without affecting the more sharply defined areas of shadow. The final PDA image was saved as a BMP file titled 20071226_0055-0135_Fraunhofer_127mmMCT×200.bmp which took up 225 KB of the SD memory card (see below) (Figure 8.3).

Ten minutes after the observing session had ended, the PDA was inserted into its desktop cradle connected to the desktop PC and *ActiveSync* was initiated. The saved BMP image file was located by browsing the PDA's contents on the desktop PC, copying it from the SD card, and pasting it into a lunar observations folder on the computer's main hard drive. The image was opened in *Corel PhotoPaint 8*, and because it was a relatively small 240 × 320 pixel image measuring 8.5 × 11.3 cm (3.3 × 4.4 in.), it was resampled to 300 dpi (but retaining its dimensions), converted to grayscale, and saved in JPEG format. The image was zoomed-in to fit the monitor, and the enhancements began with the observation still very fresh in the mind.

1. 01:00

2. 01:20

3. 01:25

4. 01:35

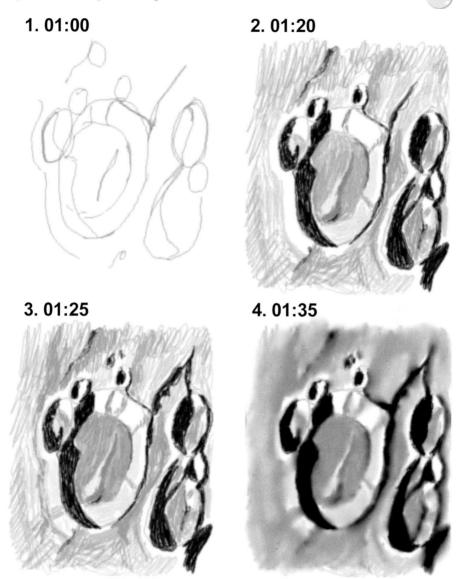

Figure 8.3. Sequence of saved field cybersketches (times given) showing the development of a freehand PDA lunar observation of Fraunhofer over 40 min. Note that the images are orientated correctly, north at top and east at right, as they were made at the eyepiece (an erecting prism was used) (photo by Peter Grego).

A combination of mouse and graphics pad input was used to perform the image enhancements. After enlarging the PDA image, a median filter was used to remove unwanted artifacts, notably pixelation. On viewing the image up close, what ought to be sharply defined areas of black lunar shadow assumed a patchiness (caused by using a pencil stroke fill rather than flood fill) and displayed a somewhat fuzzy edge. So, the first task was to sharpen the edges, which was achieved through a carefully applied smear tool and a paint brush. Shadows were intensified using a paintbrush set to darken mode; the brightest areas were brightened with the same tool. Extra coverage was then provided in the corners, using a blending tool, and finally a few extra blends and enhancements to the detail were created using the smear tool. The resulting enhanced cybersketch was faithful to the original observational drawing (Figure 8.4).

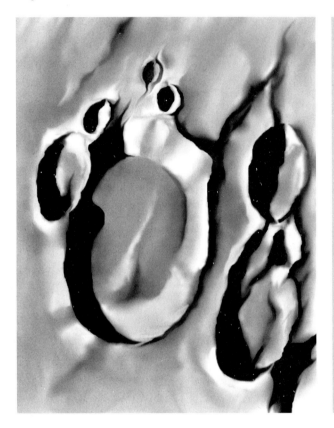

Figure 8.4. The final, enhanced observational cybersketch of Fraunhofer (Peter Grego).

Saving Cybersketches to PDA

We discussed earlier how adverse conditions can cause mobile devices to glitch (a problem experienced by many other electrical devices, including clocks and GPS systems). Some glitches are minor and require the device to be simply restarted using the hard reset button, though you'll lose any current, unsaved work in the process. More serious glitches can actually obliterate all the user's settings and remove the user's own programs installed on the PDA's main memory, wiping out a great deal of hard work. The same can happen if the PDA is accidentally dropped while in use and its battery disengages. These mishaps can quite easily happen in the dark.

PDA utilities, such as *Permanent Save* and *xBackup,* offer the facility of backing up vital data to a specified safe location such as a memory card. Anything stored on a memory card will be impervious to any dramatic system resets to the PDA's original factory configuration. System-wide backups ought to be performed every few weeks; depending on the amount of data to be backed up, the process can take anywhere between a couple of minutes to half an hour.

By saving cybersketches to the memory card, at least your hard-won observational cybersketches are safe from inadvertent deletion through system failure. It is wise to save regularly throughout an observing session, overwriting the preceding saved version of the observation, because glitches may occur at any point; if you don't have a saved version to revert to, you'll have to start again from square one. So, it's best to save your observational drawings to the PDA's removable memory card, rather than its inbuilt main memory. Most graphics programs offer the user a choice of save locations – usually defaulted to the main memory – and some programs allow these default settings to be altered and permanently set by the user.

When saving a cybersketch it is always a good idea to include the most relevant details of the cybersketch in the file name – especially the feature's name, plus the date and time of the observation. This saves having to include unnecessary text information on the cybersketch itself, and it makes things a lot easier when identifying the observations after they are transferred to the desktop PC for enhancement, filing, future reference, and sharing with other observers and organizations.

The file naming format this author generally uses is date_time (UT)_object_instrument and magnification_observer. Other details can be added, such as diagonal/erecting prism/binoviewer/filter used, seeing conditions, observing location, but the file name runs the risk of becoming inordinately lengthy after a while.

Making Freehand Field Cybersketches of Deep Sky Objects on a PDA

Everything beyond the Solar System is classed as a deep sky object (DSO). This includes double stars, open star clusters, globular clusters, and nebulae (most of those visible to the amateur lie within our own galaxy), to more distant galaxies in ever-deepening expanses of space. DSOs are pretty faint, in the main; only a couple of dozen of them are visible with the unaided eye in dark conditions. Dozens more are discernable through binoculars, while larger telescopes are capable of revealing many hundreds, even thousands, of deep sky denizens.

Before going through the process of deep sky cybersketching, a few notes about dark adaptation. To view DSOs at their best requires the eyes to become adjusted to low-light conditions. In the dark, our color vision, which relies on the cone cells in the retina, ceases to function. Instead, the retina's rod cells gradually come into play, and these are responsible for our monochromatic night vision. After about half an hour of darkness it's possible to perceive very faint objects, almost to the limit of your vision. Any bright light will instantly ruin your hard-won dark adaptation. It happens that the rods are less sensitive to red light than any other wavelengths. If the red light is bright enough for the rods to register it, but not too bright, there is minimal effect on dark adaptation, allowing a return to observing those faint celestial objects more or less immediately. To maintain dark adaptation while observing faint celestial objects, it's necessary to turn down the brightness of your PDA, UMPC, or tablet PC to acceptably low levels. Some computer programs, actually tailored for the amateur astronomer (such as the superb *DarkAdapted* for Win/Mac), do exactly this – even to the extent of turning the display's screen red. This will enable astronomical and cybersketching programs to be viewed and used without ruining dark adaptation.

Perception of very dim objects, particularly nebulae and galaxies, can be greatly enhanced using an observing technique called averted vision. Dim objects appear brighter, and their structure is more clearly discernable when the observer's gaze is directed to one side of the object, rather than directly toward it. This seemingly counterintuitive phenomenon is caused by the eye's physiology.

Two types of photosensitive cells, known as rods and cones, are found at the back of the retina; these convert light into electrical signals that the brain processes into an image. Cone cells are located in the fovea, at the center of the retina, while the rod cells are arrayed around the fovea, in the macula. Detailed color images of objects at the center of our field of view are delivered by the cones, but they require a certain degree of brightness to be activated. In dark situations, the rods are responsible for our vision; insensitive to colors, rods deliver less detailed images than cones.

To use averted vision effectively, allow some time for your pupils to dilate in the dark before observing; the larger the pupil, the more light will enter the eye, making faint objects easier to perceive.

If the DSO is too faint to see directly, try to place it within the field of a low-power eyepiece. Movement often betrays the presence of a faint object, so try moving the telescope slightly from side to side, or scan the periphery of the field. To get the best averted views, direct your gaze up to $18°$ away from the actual location of a faint object, offsetting the deep sky object toward your nose to find the 'sweet spot' of averted vision. If you're a right-eyed observer, avert your gaze to the right of the DSO; if you observe with the left eye, look to the left of the object's position.

Once your quarry is found, don't immediately move on to another faint fuzzy. Keep the object centered and spend some time using averted vision to coax out the maximum amount of detail. You may surprise yourself at just how sensitive your eyes actually are.

When drawing bright DSOs such as extensive open star clusters it is best to place the object under scrutiny at the center of the telescopic field of view. However, you might find it better to offset really faint objects to the left or right side of the field so that you can make the best of your averted vision's sweet spot.

Basic Freehand DSO Cybersketches

If you're completely new to cybersketching, unsure about using layers in a drawing, and prefer to make a cybersketch with just the basic tools in a graphics program, here are a few simple tips for producing a freehand field cybersketch of a DSO:

- Don't introduce colors into a sketch – keep it grayscale.
- Field sketches of DSOs are easier to draw in negative, i.e., black stars on a white background.
- If possible, draw a circle to represent the edge of your eyepiece's field of view and keep your sketch within its boundaries; this helps maintain your orientation while sketching.
- If the DSO is a nebulous object, draw it lightly with a light gray shade and then blur the image, or smudge it around the edges. Additional brush strokes of varying gray tones can be added on top, each tone blurred repeatedly to achieve the right look.
- Using a black round-tipped brush tool, mark the positions of the brighter stars in and around the nebula. Use a large pencil tip for the brightest, mentally using the clock system for position angle with respect to the DSO at center, along with an estimate in fractions of the radius to the field's edge. This cannot be done hastily. Take your time to place each individual bright star as accurately as you can, until you have enough field reference points to add detail and the fainter stars. It will probably suit most observers not to go into too much detail in terms of the numbers and precise positioning of the fainter stars for reasons of time. Seasoned observers usually know when they have successfully captured the essence of a particular deep sky object.

Here, now, is another observational tutorial, this time of a freehand DSO cybersketch on a PDA.

Date: August 13, 2007
Time: 01:20 UT
Feature: NGC 7331 (galaxy)
Instrument: 200 mm SCT (Meade LX90) ×75 (binoviewer, diagonal, and LPR filter used)
Cybersketching device: SPV M2000 P/PC
Cybersketching programs: *Mobile Atelier* (field sketch) and *Corel PhotoPaint* (desktop enhancement using a graphics pad and mouse)

A clear night with the Moon well out of the way, ideal for deep sky observation. After setting up a 200 mm SCT a binoviewer (a binocular eyepiece) was attached to a star diagonal, which shows a north-up, east–west flipped view. A light pollution reduction (LPR) filter was also used. Using 26 mm Plossl eyepieces with this instrument delivers a magnification of 75× and shows an actual field of view around 0.7° and an apparent field of 55° – quite sufficient for most deep sky work.

With the excellent PDA planetarium program *Astromist*, a suitable deep sky target was chosen – NGC 7331. According to *Astromist*, it's a magnitude 9.5 galaxy in Pegasus, bright and pretty large, extending 163° with a smaller, brighter middle. The galaxy was centered in the eyepiece's field of view within moments of inputting its name into the telescope's Autostar hand controller.

1. *Mobile Atelier* was used to make the field cybersketch on my SPV M2000 P/PC. In addition to dimming the PDA's display to its lowest setting, night vision was maintained by using a pre-prepared red template with a circular outline marking the field of view.

The center of the galaxy was almost stellar in appearance, and it was marked with a point at the center of the sketch. The brighter stars in the surrounding field of view were depicted first, using a black round-tipped brush, followed by progressively fainter stars using a series of smaller brush sizes. The image was saved on the PDA's SD card.

The elliptical outline of the galaxy was clearly visible. Using a smudge/blur brush, the center was blurred and extended to match the observed orientation using directed strokes to form a hazy ellipse. Another dark point was marked at its center, and this was again blurred and blended to indicate the galaxy's slightly brighter middle regions. A darker spot was then painted at the center, denoting NGC 7331's bright nucleus. To make it clear which object was observed and the orientation of the field of view, a few notes were written on the screen beneath the sketch, and the image was saved on the SD card as 20070813_0120_NGC7331_200mmSCT×75.bmp (Figure 8.5).

1. Field cybersketch

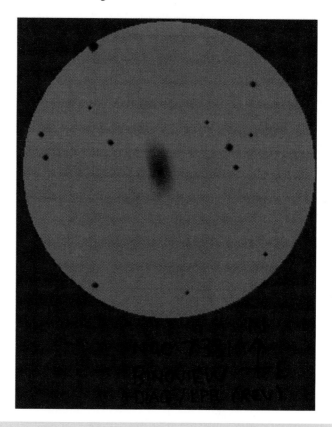

Figure 8.5. An example of a freehand field cybersketch of a DSO made on a PDA using basic graphics tools – an observation of the galaxy NGC 7331.

2. After transferring the field cybersketch image file to the desktop computer via ActiveSync, it was enhanced using the program *Corel PhotoPaint 8*. After converting this to JPEG format, grayscaling, and resampling it to 300 dpi, the circular field was cropped and pasted onto a white background.
3. Because the field observation was made on a red background, the general tone of the sky after conversion to grayscale now appeared gray. Using the tone curve tool, the gray levels were adjusted so that the gray background became white, taking care not to adjust the levels beyond this to the point where the grays of the detail in the galaxy were affected.
4. The image was then inverted so that the stars became white points on a black background and the central galactic smudges assumed shades of gray, brightening slightly toward the center.
5. Next, the image was flipped from left to right so that celestial west lay to the right and celestial north was retained at top. At this stage, further enhancements can be made by sharpening the stars, achievable by gamma adjustment (taking care to mask out the galaxy itself so that its tonal range is unaffected) and a median filter to remove pixelation artifacts. It is also possible (as in this case) to redraw the stars using a sharp-edged round brush, being careful to retain their relative sizes and positions as true to the original observation as possible (Figures 8.6 and 8.7).

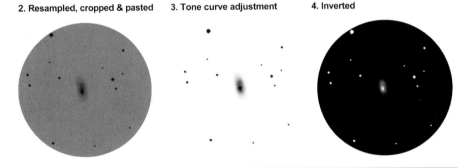

2. Resampled, cropped & pasted   3. Tone curve adjustment   4. Inverted

Figure 8.6. Stages in processing the cybersketch (photo by Peter Grego).

Following is another observational tutorial on making a freehand DSO cybersketch on a PDA using two layers. An additional layer showing nebulosity can be overlain onto a star field sketch, keeping the background in full view without altering it. The two layers can then be merged to create a combined image.

Date: July 24, 2006
Time: 00:45 UT
Feature: M6 (open star cluster)
Instrument: 15 × 70 binoculars
Cybersketching device: SPV M2000 P/PC
Cybersketching programs: *Mobile Atelier* (field sketch)

5. Enhanced cybersketch

NGC 7331

N
. W

2007 August 13   01:20 UT
Seeing All   Limiting magnitude 5
200 mm SCT x75 (with binoviewer)
LPR filter
Peter Grego (Rednal, UK)

Figure 8.7. The completed cybersketch of NGC 7331. Photo by Peter Grego.

Large binoculars are capable of revealing some truly stunning large-scale celestial sights. M6, a bright open cluster in the constellation of Scorpius, was selected from a list of suitable objects presented by the PDA program *Astromist*. According to the program, it shines at a combined magnitude of 4.2, most of its stars ranging from the 7th to the 10th magnitude. The cluster was quite easy to locate using steadily held binoculars, but, of course, full resolution of all its constituent stars wasn't possible. Instead, the cluster's brighter stars shone out from an elliptical haze filling a good portion of the 4.4° binocular field of view.

1. After centering M6 in the field of view, *Mobile Atelier* was used to make the field cybersketch on an SPV M2000 PDA. First the background stars were depicted, beginning with larger black dots for the brightest stars in the field and using these as guides for placing smaller dots for the fainter stars.
2. A layer was added to the cybersketch, enabling the hazy oval patch in which the cluster's stars were embedded (actually unresolved stars) to be depicted without affecting the image behind. The haze was depicted by painting in an oval shape with a light gray paintbrush and then blurring it. The image was then combined and saved as 20060724_0045_M6_ 15x70.bmp.

3. The observational cybersketch was transferred from the PDA's SD card to a DSO observations folder on the desktop PC's hard drive by dragging and dropping the file via *ActiveSync*. The image was resampled at 300 dpi, then converted to grayscale and JPEG format. A small amount of median filter was applied to remove some of the original's pixelation, and unsharp masking was used to make the stars' edges a little sharper (Figure 8.8).

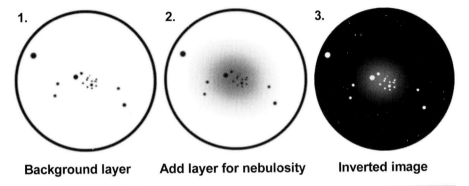

Background layer   **Add layer for nebulosity**   **Inverted image**

Figure 8.8. An additional layer was used to depict nebulosity (actually unresolved stars) in the star cluster M6 (photo by Peter Grego).

Jupiter on a PDA

Of all the planets visible through the telescope eyepiece, Jupiter has the most going on. The giant planet's disk always appears larger than 30 arc seconds across, and at opposition it can grow as large as 50 arc seconds in apparent diameter – easily big enough to be discerned as a distinct disk through steadily held 10 × 50 binoculars. Jupiter's four largest moons – Io, Europa, Ganymede, and Callisto – are bright and easy to spot through binoculars.

A small telescope will reveal a few of Jupiter's darker cloud belts and brighter zones, which appear stretched out and parallel to the equator because of the planet's rapid spin of less than 10 hours. The planet's rapid spin rate also causes its equator to bulge, producing a noticeably elliptical disk. Various features embedded within the belts and zones, festoons, prominent bright and dark spots, and the like are fascinating to view and record through telescopes larger than 100 mm. The belts and zones appear to vary in color and intensity from year to year, but the most prominent of Jupiter's cloud features are usually the north and south equatorial belts. Features within the cloud belts and zones change in appearance virtually on a daily basis, as spots, ovals, and festoons develop, drift in longitude, interact with one another, and fade away. Features always remain within their own belt or zone, and they seldom drift much in latitude.

Whole disk drawings of Jupiter depicting the relative broadness, positions, and intensities of the planet's belts and zones, plus the finer detail discernable within them, make fine snapshots of the planet's appearance to the observer at any one moment in time. Through a telescope of 200 mm or larger, an experienced observer might see so much detail on the planet during periods of good seeing that it may be difficult to depict satisfactorily.

Features on Jupiter appear to fade slightly toward the limb; some observers ignore this fading, some represent it with slight shading around the edge, while others choose to simply draw the belts, zones, and any features near the limb slightly less distinctly than the rest – it's a matter of personal preference.

After drawing in the features that are readily discernable, the observer is free to concentrate on one small, specific area at a time, attempting to tease out fine detail as seeing conditions allow. Detail requires some experience to see and fully appreciate, and an inexperienced observer may not see much more than the main belts and zones crossing Jupiter's disc. Familiar with the appearance of Jupiter, an experienced observer is able to spot, in an instant, the emergence of a vague feature at the planet's limb, while the same feature may not even be visible to an inexperienced observer when it is in full view on the planet's central meridian. Experienced observers take advantage of moments of good seeing, and they tend not allow the eye to wander around.

Here, now, is an observational tutorial of a freehand cybersketch of Jupiter on a PDA using two layers.

Date: April 3, 2005
Time: 23:45 UT
Feature: Jupiter
Instrument: 200 mm SCT (Meade LX90) ×285 (star diagonal used)
Cybersketching device: Jornada 540 P/PC
Cybersketching programs: *Mobile Atelier* (field sketch) and *Corel PhotoPaint* (desktop enhancement using a graphics pad and mouse)

1. After centering Jupiter in the eyepiece, a few moments were taken to absorb the overall scene to discern what large or particularly prominent features were arrayed across the entire Jovian disk. A black background was selected, upon which a white-filled ellipse was drawn to approximate the oblateness of Jupiter.
2. A layer was added, upon which to draw the planet's features without affecting the elliptical blank. For speed, it was decided to make this a grayscale drawing, rather than attempting to depict colors. The darkest belts were drawn first, using a broad faded-edge gray brush, no detail added. Other belts and zones were added in lighter shades of gray, leaving the brighter areas white.
3. A smaller brush was used to add the detail in various tones of gray. Most of the detail in this particular observation lay between the north equatorial belt and the equatorial zone.
4. A blurred brush was used to blend and smudge features.

5. More sharply defined areas of detail were zoomed-in and enhanced using a fine brush. The image was saved on the PDA's CF card as 20050403_2345_Jupiter_200mmSCT×285.bmp.
6. After placing the PDA in its desktop cradle, the observation was transferred from the CF card to the desktop PC via *ActiveSync*. The cybersketch was resampled to 300 dpi, converted to grayscale, and saved as a JPEG file. The coarse pixelation to the details and around the planet's edge was removed effectively using a median filter.

Colored Double Stars on a PDA

One of the advantages of using a mobile computer to make astronomical cybersketches is the ability to introduce color into a drawing. Deep sky work doesn't appear to contain a great amount of color because most DSOs are too dim to trigger the eye's color receptors. But there is one area of deep sky work in which color is perceived, often in a striking manner – colored double star observation (Figure 8.9).

When making a field cybersketch of colored double stars, a positive drawing (bright stars on a black background) is the easiest method. If a negative drawing is made, then the colors will be transformed into their spectral opposites when the image is inverted during processing – red appearing blue, blue appearing red. There are several ways of getting around this, but the method described here is the simplest.

The perceived color of a star can be portrayed with a dot of the appropriate color and size. People's perception of color varies, so we're looking at a general impression more than an exact science. If you choose to exaggerate the colors you perceive, that's perfectly fine. By blurring the brighter stars during the enhancement process and then adding a further bright white dot of the right size at its center, the impression of slight stellar glare can be created, which enhances the look of the cybersketch. Note that bright-colored stars should also be depicted with a white central white dot – not a colored dot (Figure 8.10).

Observational Cybersketching on a Tablet PC/UMPC

Tablet PCs and UMPCs may have the advantage of screen size over PDAs, but they are heavier and more tiring to hold over long periods of time. Strapping big folk who work out with dumbbells may be able to hold a tablet PC in one hand throughout an entire observing session without giving it a second thought, but most people may feel slight discomfort after a while if they remain standing while holding the device in the hand. It would be nice if manufacturers incorporated a clip-on hand strap to the back of their tablet PCs (a bit like the soft hand cradle featured on camcorders) so that they can be held up or lowered to the side of the body without needing to grip it between

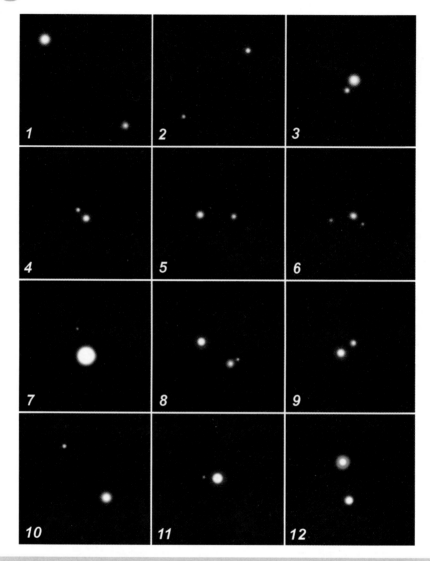

Figure 8.9. A selection of the sky's most beautiful colored double stars (shown as seen through an astronomical telescope with north at top). 1 – Beta Cygni (Albireo), 2 – Eta Persei, 3 – Epsilon Boötis, 4 – Xi Boötis, 5 – Gamma Delphini, 6 – Iota Cassiopeiae, 7 – Beta Orionis (Rigel), 8 – Upsilon Andromedae, 9 – Alpha Herculis, 10 – Alpha Canum Venaticorum (Cor Caroli), 11 – Alpha Scorpii (Antares), 12 – Beta Scorpii (photo by Peter Grego).

the thumb and fingers. Homemade solutions are, of course, possible. If you're handy at DIY, a hand strap may be made, or perhaps the tablet PC could be held on a platform against the abdomen using a neck strap. Alternatively, an adjustable platform can easily be adapted to fit on a monitor bracket and fixed to the telescope pier, or a tripod can be modified to hold the tablet PC. A table may suit you just fine, though – as long as you don't have to stoop too far to make the cybersketch.

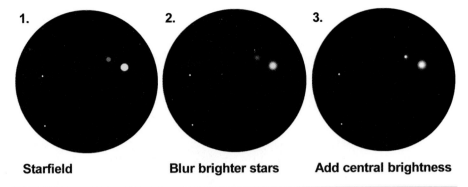

1.   **2.**   **3.**

Starfield   **Blur brighter stars**   **Add central brightness**

Figure 8.10. Stages in making a cybersketch of a colored double star. This observation of Albireo in Cygnus, a famous colored double with a golden primary of magnitude 3.1 and its steely blue companion of magnitude 5.1, was made using a 102 mm refractor – the field of view is about 20 arc minutes across (photo by Peter Grego).

As discussed above, initial field drawings on large-screen devices, in which the subject's main features are depicted, might not benefit from being made at maximum size to fit the screen. More accuracy is possible if the cybersketcher can take in, at a glance, the whole area bounding the image being drawn, without having to scan the eyes over a wider area than is necessary (Figure 8.11).

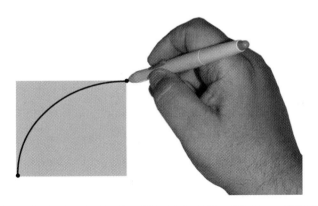

Figure 8.11. Determining the optimum size for making field cybersketches on UMPCs and tablet PCs. Detail within the cybersketch can be added by enlarging the image on-screen by zooming in (photo by Peter Grego).

UMPCs and tablet PCs offer a much greater degree of brightness and contrast control than PDAs, so it's possible to adjust the screen's gamma settings to a level suitable for deep sky work requiring dark-adapted eyes; there are a number of programs available that are specially intended for astronomical use, including the program *DarkAdapted* (see above), which is able to set the screen to night vision-friendly tones of deep red.

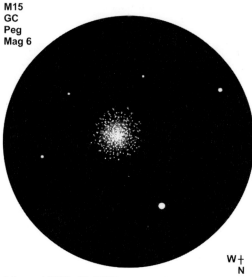

M15
GC
Peg
Mag 6

W +
N

8 August 2005, 00:45 UT
Seeing AII, clear and windless conditions
125mm MCT (ETX-125) x95, no filter
Observation by Peter Grego, Rednal, UK

M15 clearly resolved, a lovely sight with a relatively
compact nuclear condensation of stars.

Figure 8.12. Globular cluster M15, observed on August 8, 2005, through a 127 mm MCT and sketched with a tablet PC (photo by Peter Grego).

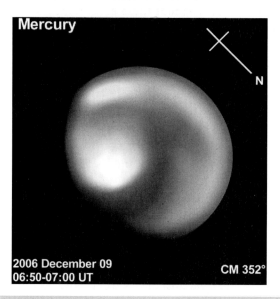

Mercury

N

2006 December 09
06:50-07:00 UT   CM 352°

Figure 8.13. Observational cybersketch of Mercury made on December 9, 2006 (photo by Peter Grego).

What follows is an observational tutorial of a freehand field cybersketch of a lunar feature object on tablet PC.

Date: July 23, 2008
Time: 02:15-03:00 UT
Feature: Plinius
Instrument: 200 mm SCT (Meade LX90) ×200 (star diagonal and binoviewer used)
Cybersketching device: ViewSonic Tablet PC V1100
Cybersketching programs: *Corel PhotoPaint*

Among the features visible near the terminator of the 20.2-day-old waning gibbous Moon, the crater Plinius (43 km in diameter) caught my attention. I had previously observed this prominent crater several times, under morning and late afternoon illumination, but had never before made an observational drawing of it with such an advanced evening shadow.

1. Using a ViewSonic Tablet PC V1100, a 300 dpi grayscale JPEG image was created with the dimensions of 10 × 10 cm (4 × 4 in.), with a plain white background. Using a light gray 5 pt art brush, the outlines of the main features of Plinius and its immediate environs were depicted with loose strokes.
2. The dark, shadow-filled areas were outlined with a black brush.
3. These broader black areas were completed using the fill tool.
4. Further detail and shading in and around Plinius was added using a brush and several shades of gray. Lighter gray areas were applied on an 'if darker' setting to preserve preexisting lines of detail. Darker grays were shaded loosely. Throughout these early stages the drawing was kept loose, without much concern over the jumbled appearance of brush strokes.
5. Using a dither brush, coarser features of tone were blended to produce areas of more even, flowing tone. Dithering was performed using 'if lighter' and 'if darker' settings to preserve the more distinct parts showing solid detail.
6. A smear brush was used to produce further smoothing and blending effects.
7. A light median filter was used to smooth things out a little further.
8. Finally, the tone curve was adjusted, and the cybersketch was cropped and saved on the tablet computer's hard drive as 20080723_0215-0300_Plinius_200mmSCT×200.jpg. In addition to saving it on the device, the image was sent as an e-mail attachment directly to my own e-mail address directly from the field using the device's WiFi connection to my home router.
9. Once indoors, the image was retrieved from my e-mail in-box on the desktop computer and opened in Corel PhotoPaint 8. The paper size was changed to A5 format, and the full observational details were added, completing a successful cybersketch (Figure 8.14).

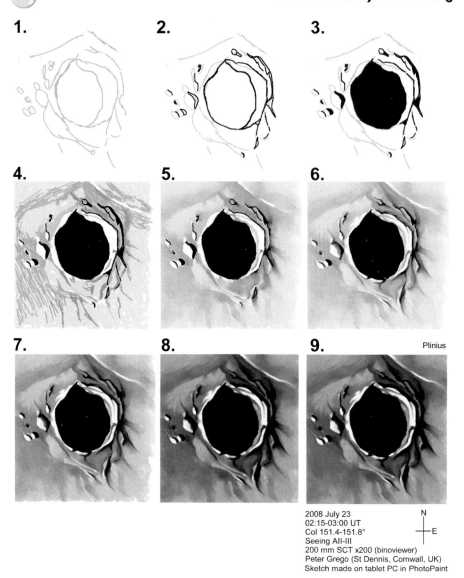

Figure 8.14. Stages in the making of a freehand observational cybersketch of the lunar crater Plinius on tablet PC (photo by Peter Grego).

Plinius

Plinius

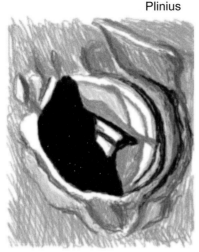

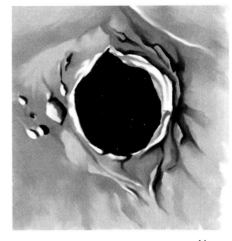

2005 November 21
04:47-05:17 UT
Col 135.5-135.7°
Seeing AII-III
200 mm SCT x300
Peter Grego (Rednal, UK)
Sketch made on PDA in Mobile Atelier

2008 July 23
02:15-03:00 UT
Col 151.4-151.8°
Seeing AII-III
200 mm SCT x200 (binoviewer)
Peter Grego (St Dennis, Cornwall, UK)
Sketch made on tablet PC in PhotoPaint

N
E

Figure 8.15. A comparison of freehand observational cybersketches of the same lunar feature – Plinius at an evening illumination – made at the eyepiece using a PDA (Mobile Atelier) and a tablet PC (Corel PhotoPaint). The lower resolution of the PDA is evident, although further work, either at the eyepiece or on the PC indoors, will undoubtedly greatly improve its appearance (photo by Peter Grego).

Observational Drawing Aided by Digital Images

Ever since amateur astronomers began to explore the vast possibilities of the CCD chip, digital images have often been blamed for the decline of traditional observational drawing. However, CCD images are far from being the astronomical sketcher's nemesis – on the contrary, they serve as an inspiration to visual observing. CCD images can suggest new features to look for through the eyepiece, be they previously unknown lunar albedo or topographical features, faint details on the planets, or certain aspects of the deep sky previously unsuspected by the observer.

Digital images can form an integral part of the cybersketching process in several ways: simultaneous sketching (displaying a real-time image on a computer in a window adjacent to the cybersketch window), sketching from live or recorded images (i.e., from a TV monitor or a computer screen), or drawing directly onto near-live low-contrast images (which can either be real or computer generated). Some of these techniques, of course, may be considered 'cheating' by some folk who have no intention of ever setting down their graphic endeavors in anything but traditional pencil and ink.

Simultaneous Cybersketching

It's no problem to simultaneously display several images on-screen using the 'windows-based' GUI of Windows PC or Mac computers. In simultaneous cybersketching, one window shows a live webcam image of an astronomical object, and this image is copied from directly into an adjacent graphics program window. This technique removes the need to look through the telescope eyepiece; the webcam itself serves as an electronic eyepiece. Because the live image is subject to the whims of atmospheric seeing, cloud coverage, and so on, the experience of simultaneous sketching is similar to observing at the eyepiece. It's just as visually challenging as at-the-eyepiece work, and if a suitable extension cable or wireless link is set up, it can be performed remotely from the telescope, in the comfort of the house.

Sketching from Live or Recorded Images

While simultaneous sketching involves using the same computer monitor to display a live CCD image and an observational cybersketch in separate windows, live or recorded CCD images displayed on other devices can, of course, be copied from in order to produce an observational sketch (Figure 8.16). For example, digital video footage of a lunar scene can be played in loop mode and used as a basis for a drawing, either a cybersketch or using traditional pencil and paper. The fact that the live or recorded image will display inconsistencies due to varying seeing conditions simulates the experience of being at the eyepiece. The sketcher is still using observing skills to perceive finer detail in moments of better seeing.

Live Deep Sky Sketching Using a Video Camera

Dale Holt (Chipping, UK), an active visual observer and excellent 'traditional' sketcher, is one of the most active proponents of deep sky sketching based on live CCD images. Let's take a look at the techniques that Dale uses, in his own words:

> 'I am a committed visual observer – apart, that is, from one transgression. Ever since I first tried out a Mintron deep sky video camera some four years ago I have enjoyed the benefits these video cameras can offer the deep sky observer.

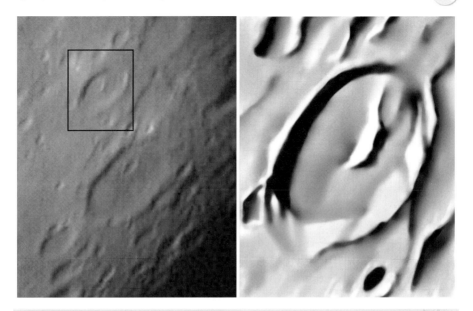

Figure 8.16. An observational example of a lunar cybersketch on PDA based on recorded video footage. Based on a 2-min-long AVI sequence taken on November 18, 2005, at 01:30 UT using a 200 mm SCT (Meade LX90) and Philips ToUcam PCVC740K webcam, the author made a sketch of a small area contained in this footage – the crater Stevinus – using a Jornada 540 P/PC with the program *Mobile Atelier* (photo by Peter Grego).

Figure 8.17. Dale Holt made this observational sketch of the globular cluster M3 on April 6, 2008 at 22:00 UT. He based the image on a live monitor display taken through a 350 mm Newtonian and Watec 120 N video camera (photo courtesy of Dale Holt).

Figure 8.18. Dale Holt made this observational sketch of the famous Ring Nebula M57 and the fainter background galaxy IC1296 on May 6, 2008 at 00:05 UT. He based the image on a live monitor display taken through a 350 mm Newtonian and Watec 120 N video camera (photo courtesy of Dale Holt).

Soon after seeing the Mintron 121 V in action I purchased one for myself. This camera showed me things virtually in real time that I had never seen before. Structure in galaxies way beyond my telescope's 14 inches of aperture was apparent; the central star of M57 became easy, even when using a 6-inch refractor; and the Eskimo Nebula really did look like a native of far northern latitudes, with his face sheltered deep within a fur fringed hood!

It wasn't long before I upgraded to a Watec 120 N camera, a smaller and more sensitive unit than the Mintron. This meant that I could go much deeper picking out deep sky objects that virtually no amateur could see visually (at least not under my skies of limiting magnitude 5).

It was only a matter of time before I started to combine the use of the Watec camera with another interest of mine, sketching. I am lucky to have an office (or, as some call it, a 'warm room') built onto the side of my run-off observatory. By running cables from the observatory through some buried plastic pipe into the office I can view objects captured via my 14-inch f/5 Newtonian and Watec camera combination on an 11-inch black and white security monitor in luxurious comfort.

Using pencils and pastels I can draw what I see easily and accurately without discomfort and with an adequate level of background lighting. It is usually at this point I hear cries of 'cheat,' and to a certain extent I can understand such comments. I, too, have felt more than a pang of guilt when comparing my sketches with those of other observers' work, so hard earned at the eyepiece in freezing conditions and by the red glimmer of a dimmed torch.

But I endeavour to put such thoughts behind me because this style of observing and capturing hand drawn images offers the best of both worlds. It embraces the CCD chip of today, so revered by astro-imagers, and it also follows in the traditions of night sky sketchers, started by Galileo Galilei.

The work I produce with my pencils is a personal rendition of what I see on the screen – it is not a perfect copy, but it is personal and unique. The image I'm sketching improves and deteriorates with seeing just as if I was glued to my 22 Nagler. The biggest differences being that my eyepiece is now 11 inches across and the detail visible beats what I can see with my 20-inch Dobsonian, hands down. In fact, by my reckoning it is likely to surpass what a 25-inch telescope would deliver to my retina from my home location!

I concede that the magic of eyepiece observing is missing, and I wouldn't want to give that sensation up, but this way of observing and sketching certainly has its merits and I'm proud to be combining the old with the new and capturing those deep sky wonders so dear to us with CCD chip and pencil.'

Using Digital Image Templates to Make Lunar Observational Drawings

The Moon is the most detailed object in the skies and by far the most difficult of celestial objects to portray accurately in an observational sketch. Aside from the problem of depicting the range of gray tones and

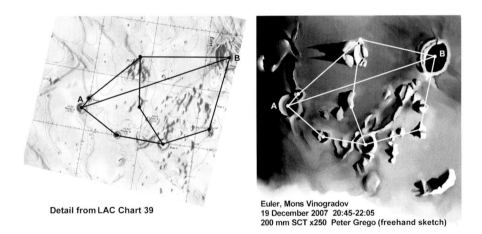

Detail from LAC Chart 39

Euler, Mons Vinogradov
19 December 2007 20:45-22:05
200 mm SCT x250 Peter Grego (freehand sketch)

Figure 8.19. A freehand cybersketch of the Euler and Mons Vinogradov area of the Moon, made on December 19, 2007, from 20:45 to 22:05 UT using a 200 mm SCT ×250 and ViewSonic Tablet PC. To demonstrate the positional accuracy of this freehand depiction, the sketch is compared with a detail from LAC 39 (Lunar Astronautical Chart series) showing the area. Both are matched to the baseline A–B (the center of the crater Brayley B to the center of Euler) and common reference points are indicated (photo by Peter Grego).

all the fine detail, the observer needs to get the relative positions of the features as close to their actual configuration as possible if the drawing is to work out right, especially if there's a lot of widespread, disparate detail on view (Figure 8.19).

Thankfully, there is a modern solution to the problem of attaining the positional accuracy of lunar observational sketches, somewhat akin to the photographic techniques pioneered by Johann Krieger in the late nineteenth century (see above). CCD chips are so sensitive that even a basic, entry-level compact digital camera, held to the telescope eyepiece, is capable of capturing a reasonably good image of the Moon. Such digital images, even if they are not terribly good, can be used as templates for general lunar observational sketches – not only cybersketches, but traditional pencil and paper sketches, too, as explained below.

Observational Lunar Sketches Based on a Near-Live Digital Image Template

To ensure some degree of accuracy, many traditional pencil and paper visual lunar observers select a feature to observe in advance of the observing session itself and pre-prepare an outline blank. If a feature is selected at the eyepiece, an outline drawing showing the main topographic features may be prepared indoors. Making a general outline of the main features prior to observing permits the observer to concentrate on depicting the subtle features and attend to detail, instead of expending excessive amounts of time and effort in attaining positional accuracy. However, this still requires an accurate outline map to begin with, and moreover it may fail to address the problems of illumination and libration, the latter of which can distort lunar features considerably, especially those near the limb.

Often drawings of smaller, less complex features can be made at the eyepiece without reference to a map at all. Still, either technique is rather involved, requiring great concentration on the part of the observer. Unless carefully drawn reference points are used, the most minor but carelessly placed detail can throw out the overall accuracy of an observational drawing.

The CCD camera offers a neat technological solution to the problem of attaining good visual observational sketching accuracy. Using nothing more exotic than a compact digital camera and afocal imaging through a telescope, images of the Moon can be secured in seconds. A high magnification sweep of the terminator is obtainable in a dozen or so overlapping frames.

After transferring the images to the desktop PC, they are viewed in order to target a particular feature for detailed scrutiny at the eyepiece. The image is then cropped to the desired extent and enlarged to a scale suitable for drawing (Figures 8.20, 8.21, 8.22, 8.23 and 8.24). The image is converted to grayscale and printed out on quality smooth drawing paper – the image's

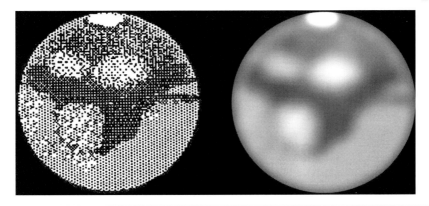

Figure 8.20. It is possible to revisit blocky old drawings made on primitive graphics programs and enhance them using modern graphics software to produce very nice results with relatively little effort. Here, a simple black and white observational drawing of Mars, made in 1988 on an Apple Macintosh, has been enhanced to produce a fairly convincing grayscale image. This involved using a tolerable level of Gaussian blur (removing the dot artifacts but retaining the broader details), brightening the south polar cap, and finally applying a median filter to smooth the edges. The image shows Mars, observed by the author on September 20, 1988, through a 100 mm (4-in.) Wray achromatic refractor. The planet's central meridian is 283°; the dusky tract of Syrtis Major lies just to the right of center, and the brighter Hellas lies above (to its south) (photo by Peter Grego).

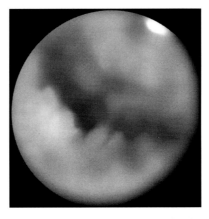

Figure 8.21. This observational drawing of Mars, coarsely cybersketched on an old Apple Macintosh back in 1988, has been given new life thanks to digital enhancement (photo by Peter Grego).

brightness and contrast levels have been adjusted to low levels while keeping the main features discernable, so that a pencil sketch made over the template will be as clear as possible. Although low-resolution digicam images can be used, higher resolution CCD images will be more clearly defined. Of course,

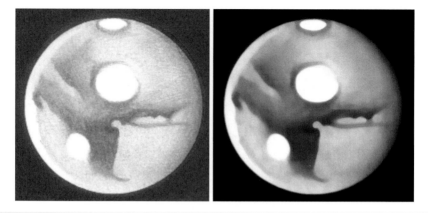

Figure 8.22. This observational pencil drawing of Mars, made on October 30, 1988, at 00:15 UT, has been scanned and enhanced in a graphics program (photo by Peter Grego).

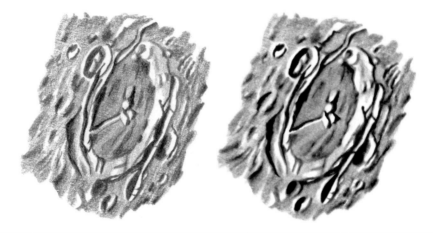

Figure 8.23. This observational pencil drawing of the lunar crater Petavius, made on November 21, 1983, has been scanned and enhanced in a graphics program (photo by Peter Grego).

too much detail in a CCD image template is undesirable; the aim is to get the relative positions of the main features right so that the only fine detail are features that have actually been observed through the eyepiece.

After returning to the telescope eyepiece, the detailed visual study of the area in question is commenced. Of immediate advantage is the ability to draw relatively large and complex areas of the lunar surface with surprising speed and accuracy. Confident, armed with the knowledge that the big picture has been attended to, the observer is free to concentrate on fine areas of detail – detail that might otherwise have been neglected in a struggle to achieve positional accuracy (Figure 8.25).

Figure 8.24. An observational sketch of the beautiful comet Ikeya-Zhang, made on May 3, 2002, was scanned and converted into this cybersketch (photo by Peter Grego).

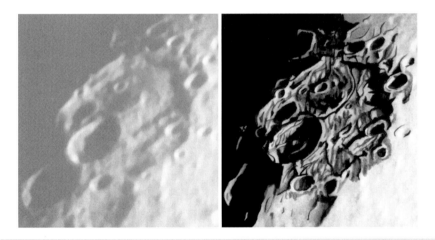

Figure 8.25. The author's first attempt at using a digital image template as the basis for a traditional pencil lunar observational sketch. Centered on the large crater Janssen, the observation was made on March 23, 2000, using a 250 mm Newtonian ×250 and Casio QV-11 compact digicam (photo by Peter Grego).

Following is an observational tutorial showing how to make a field cybersketch of a lunar feature on a tablet PC based on a near-live digital image template.

Date: September 19, 2005
Time: 04:00–04:30 UT

Feature: Gauss
Instrument: 127 mm MCT (Meade ETX-125) ×140 and Olympus E300 DSLR
Cybersketching device: Hammerhead HH3 tough tablet PC
Cybersketching program: *Corel PhotoPaint*

One of the Moon's largest craters, the 177 km diameter Gauss lies quite close to the Moon's northeastern edge. Being within the libration zone, its apparent shape and proximity to the limb varies with the whims of the Moon's monthly libratory 'wobbles.' Good combinations of illumination (which, by virtue of its location, must be an evening illumination to be seen well) and libration don't happen too often, but this morning presented a reasonably good opportunity to admire Gauss in all its glory.

Numerous digital images of the Moon were secured with a DSLR at the prime focus of a 127 mm CT and saved on the camera's CF card. The images were transferred to the tablet PC through a USB-connected flash card adapter. The Moon's terminator was then studied to select the best candidate for an observational drawing, bearing in mind that the chosen feature needed to be fairly large if the digital image was to be used as an observational template. Gauss was the ideal choice.

1. On the tablet PC, a suitable image was selected from those just taken, opened in *PhotoPaint* and cropped to size. Once resampled to 300 dpi, resized, and converted to grayscale, this image was used as the template for an observational cybersketch.
2. Black shadowed areas were more clearly defined and filled.
3. General detail in the area was outlined using a gray brush.
4. Brighter areas reflecting sunlight were added with a white/light gray brush.
5. Finally, smudge, smear, and undithering tools were used to smooth, blend, and define the observed features. The cybersketch was saved as 2005_20050919_0400-0430_Gauss_127mmMCT×140.jpg (Figure 8.26)

Here, now, is an observational tutorial describing a field cybersketch of a DSO on a tablet PC based on a near-live digital image template.

Date: October 24, 2005
Time: 01:30 UT
Feature: The Orion Nebula (M42 and M43)
Instrument: 200 mm SCT (Meade LX-90) ×140 (LRP filter) and Olympus E300 DSLR
Cybersketching device: Hammerhead HH3 tough tablet PC
Cybersketching program: *Corel PhotoPaint*

Despite the Moon being in the sky at the time of this observation, the Orion Nebula (M42) appeared an inviting target for DSLR imaging, using the image as a template to practice making a field cybersketch. The great advantage of this technique is that it achieves positional accuracy for the star field while the nebulous detail may be added or overlain as a separate layer and combined when finished.

1. Digital image template   2. Black areas added   3. Detail outlined

4. Brighter areas added   5. Smudge, smear & undither

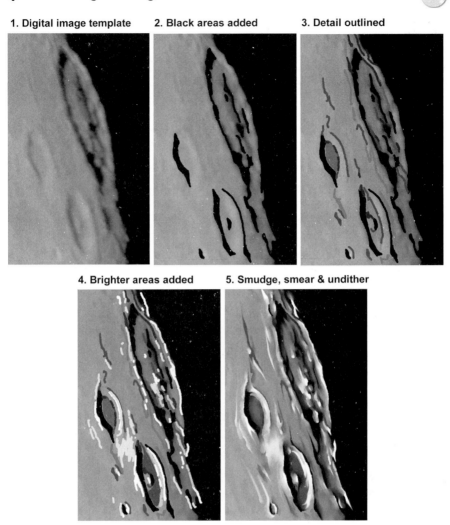

Figure 8.26. Stages in making a field cybersketch on tablet PC based on a fresh (near-live) digital image (photo by Peter Grego).

1. Using a DSLR at prime focus of the 200 mm SCT, M42 was imaged in a 30-s exposure at 400 ISO and saved on the camera's CF card. The image was transferred to tablet PC via a USB-connected flash card adapter.
2. After opening the image in *PhotoPaint*, the image was cropped and resampled to 300 dpi. Color tones were reddened, tone curves adjusted, and overall contrast lowered, reducing the visible nebulosity and the number of fainter background stars. This image was circular cropped to match the 30 arc minute

diameter visual field of view through the telescope and used as the observational cybersketching template.

3. Individual stars were painted as white dots of appropriate size over the brighter stars in the template. Since the template stars were red, the fresh brush marks were clearly visible. Once the brighter stars had been carefully depicted, the image was converted to grayscale and tone curves further adjusted so that only the fresh brush marks were visible.

4. The star field was then carefully overdrawn with nebulosity, depicting both M42 and M43 (de Mairan's Nebula) using both a brighten tool and a large smooth brush with high transparency and a soft faded edge. Finally, additional fainter background stars were added freehand with a smaller white round-tipped brush, placing them with respect to the accurately positioned brighter stars depicted earlier. The completed observational cybersketch was saved as 20051024_0130_M42_200mm SCT×100.jpg (Figure 8.27)

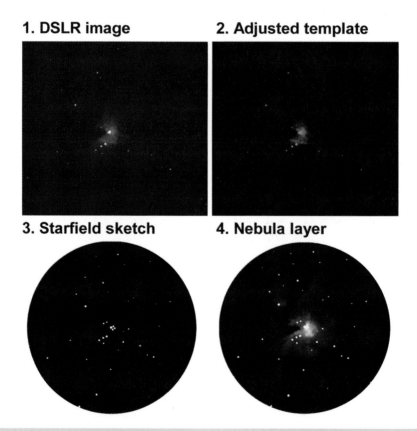

1. DSLR image   **2. Adjusted template**

3. Starfield sketch   **4. Nebula layer**

Figure 8.27. Stages in using a digital image template to make a cybersketch of M42, the Orion Nebula (photo by Peter Grego).

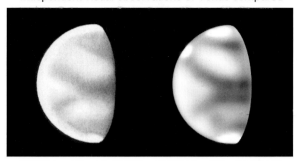

Independent visual observations of Venus compared

2007 May 07   19:30 UT   2007 May 07   20:55 UT
66mm APO refractor   127mm MCT x200
Nigel Longshaw   Peter Grego
(Chadderton, Oldham, UK) (Rednal, Birmingham, UK)

Figure 8.28. Observations of Venus compared. A traditional pencil sketch made by Nigel Longshaw and a PDA cybersketch by the author, both made on the evening of May 7, 2007 (photo by Peter Grego and Nigel Longshaw).

Digitally Revitalizing Conventional Observational Drawings

Cybersketching is not just limited to making observational drawings from scratch. Graphics programs offer the ideal set of tools to revisit and revitalize any traditional observational drawings in your archives. After being digitally imaged or scanned, observational drawings made in pencil, pastel, chalk, or ink can be manipulated and enhanced to become more faithful and better looking representations of what was actually observed. It's even possible to give new life to low-resolution material, photocopied matter, and poorly sketched observations.

Halftoning

Ink stippling is one method of reproducing observational drawings. It looks superb when it's done skillfully. Closely spaced dots of black ink are used to convey the illusion of shade; the darkness of the shade increases with the more closely spaced and/or bigger the dots are. Areas of black shadow are simply blacked in with ink and a brush. Being composed of thousands of individually applied black dots, stippled drawings reproduce very nicely even when they are photocopied. Stippling takes a great deal of time and requires patience and an exceedingly steady hand (Figure 8.29).

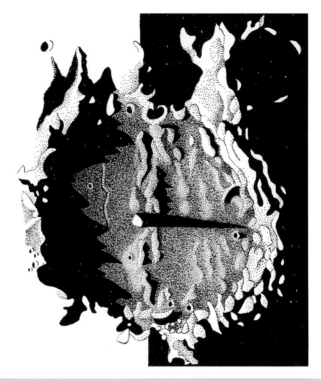

Figure 8.29. An example of an ink-stippled lunar observational drawing – the crater Alphonsus on the lunar morning terminator, observed by Phil Morgan on March 26, 2007 (photo courtesy of Phil Morgan).

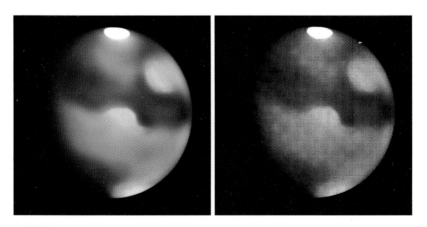

Figure 8.30. An observational cybersketch of Mars, made on August 21, 2005, at 02:25 UT using a125 mm MCT ×180, converted to halftone for photocopied reproduction (photo by Peter Grego).

One reason why stippling has been so popular among lunar (and some planetary) observers is the stippled drawing's ability to be reproduced in photocopied format without inordinate loss of information. Although pencil sketches may look great 'in the flesh' and when laser copied, standard photocopies simply don't do graphite justice, and standard photocopies are what many astronomical societies use as their method of newsletter reproduction (Figure 8.30).

Graphics programs can be used to convert scanned pencil drawings (or cybersketches) into black and white halftone images – effectively making them stippled sketches, capable of being photocopied and retaining much of their original tonal information. Halftone may be applied to any astronomical observational drawings, not just lunar and planetary work.

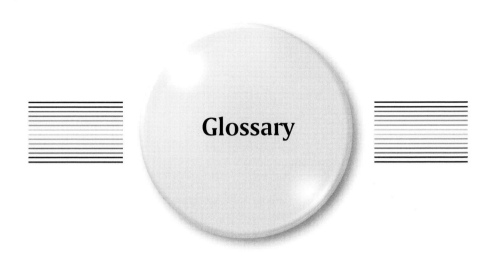

Glossary

Bit
The basic unit of computing information, with a value of either zero or one.

Bluetooth
A radio communications standard enabling devices to connect with each other to share information.

Byte
A set of 8 bits representing a single character of computer memory. File size, hard disk space, and computer memory are measured in bytes, kilobytes, gigabytes, and terabytes.

CAD
'Computer-aided design'; using precisely computed graphics to create virtual shapes and objects. Some CAD programs allow objects to be designed in three dimensions and viewed from any angle.

CCD
'Charge-coupled device'; a tiny chip made up of an array of tiny light-sensitive pixels and used in digicams, camcorders, webcams, and other electronic imaging devices.

CD
'Compact disk'; a data storage medium consisting of a circular disk that is read by a laser. CDs can hold up to 700 MB of data.

CF
'Compact flash'; a solid-state memory card used in older PDAs and digital cameras.

CGI
'Computer-generated imagery'; most often used to refer to computer-animated images.

Computer
A device capable of executing a program of instructions. Computers can be mechanical, biological, or electronic devices.

CPU
'Central processing unit'; commonly referred to as a computer's 'processor.' It's the core of a computer's 'brain' that processes every computational task. The higher a processor's 'clock speed,' given in megahertz (MHz) or gigahertz (GHz), the faster and more efficient a computer will be; 1 MHz equals one million cycles per second, while 1 GHz equals one billion cycles per second (1000 MHz). Each computer instruction requires a fixed number of cycles, so the clock speed determines how many instructions the processor can execute each second.

CRT
'Cathode ray tube'; used in old-style computer monitors, these displays used electron beams to rapidly scan across a phosphor screen to produce images.

Cybersketch
An electronic drawing.

DSO
'Deep sky object'; any body found outside the Solar System.

DVD
'Digital versatile disk'; a data storage disk of the same size as a CD but capable of holding up to 4.7 GB of data.

Ephemeris
A calculated table of astronomical data.

Field drawing
An observational drawing made at the eyepiece, or 'in the field.'

Flash memory
A non-volatile solid-state form of memory.

Floppy disk
An old form of magnetic data storage.

Freehand
A drawing made without any graphic aid.

GPS
'Global positioning system'; portable devices equipped with GPS can pinpoint their exact geographical location by connecting to a network of satellites in Earth orbit.

Graphics file format
The way in which image data have been stored. Common graphics formats accessible by most graphics programs include JPG, TIFF, GIF, PNG, and BMP. Other formats may be proprietary and can only be opened within the program used to create them.

Graphics tablet
A computer peripheral that enables input via a flat electronically sensitive surface and a stylus, used mainly within drawing and graphics programs to produce and manipulate images.

Graphics program
A program enabling the creation, manipulation, and saving of digital drawings, photographs, and images in various formats.

GUI
'Graphical user interface'; a virtual desktop containing icons that, when clicked upon, open up files and programs in rectangular windows.

HDD
See **Hard drive.**

Handheld PC (H/PC)
A type of miniature laptop computer, also called a palmtop.

Hard disk drive
See **Hard Drive.**

Hard drive (HD/HDD)
A data storage device where programs, files, and folders are physically located. Hard drives usually consist of a stack of rapidly spinning disks upon which data are encoded magnetically, mounted inside a sealed box.

H/PC
See **Handheld PC.**

Keyboard
The primary text input device for most types of computer. Letters are usually arranged in the traditional QWERTY fashion. Most PDAs have the facility to display a virtual keyboard on-screen, while a few PDAs actually have small, physical keyboards to facilitate text input.

Laptop
A laptop computer, often called a notebook computer, is a small, portable personal computer that may also be battery powered. From small to large, laptops come in the following categories: subnotebook, ultraportable, notebook, and large laptop. The average laptop typically weighs less than 2 kg and is thinner than 75 mm.

LCD
'Liquid crystal display'; thin displays used in computer monitors, laptops, tablet PCs, and PDAs. TFT-LCDs (thin film transistor LCDs) have active matrices, each sub-pixel being controlled by its own transistor. These are brighter and have more contrast than older passive matrix LCDs and offer a wider range of viewing angles.

LED
'Light-emitting diode.'

Monitor
A visual display unit. Old-style CRT monitors have now been largely replaced by lighter, clearer, lower power-consuming LCD monitors.

Moore's law
Based on the fact that computer circuitry is becoming increasingly smaller, Moore's law predicts that the processing power of computers should double every 18 months or so.

Motherboard
The computer's main circuit board, into which are plugged the CPU, RAM modules, expansion cards, and other components.

Mouse
A small, hand-operated device used on a flat surface in order to move an on-screen cursor and to select actions. Some mice are physically connected to the computer via a wire; others

are wireless and offer more freedom. Old-style mice used a physical ball to translate hand movement into cursor movement, while the latest ones use a low-power laser to determine movements more precisely.

Operating system (OS)
A computer program, first to operate on starting up a computer, which provides the environment in which all other programs may function. Most computers have a Windows-based OS. Alternatives include Linux, Mac.

Orrery
A mechanical device that simulates the orbits of the planets and their satellites around the Sun.

OS
See **Operating system.**

Palmtop
See **Handheld PC.**

PC
See **Personal computer.**

Personal computer (PC)
A home computer, commonly used in reference to an IBM-compatible computer with a Windows-based operating system. The term PC is not usually used when referring to Mac OS computers.

Pixel
A 'picture element'; images on computer screens are made up of many thousands of these tiny dots.

PDA
'Personal digital assistant'; a handheld computer.

PIM
'Personal information manager'; a handheld organizer.

Planisphere
A physical device (or computer program) that displays the position of the stars for any given date and time.

Pocket PC (P/PC)
A PDA that fits on the palm of the hand or in a pocket.

P/PC
See **Pocket PC.**

Printer
A peripheral device that produces a paper hard copy of an image, either in color or black and white. Inkjet printers use liquid inks, while laser printers use toner to produce an image.

RAM
'Random access memory'. Memory chips connected to the computer's motherboard store program information retrieved from the hard drive when a program is opened, since data can be read from RAM faster than it can be read from the hard drive. Increasing a computer's RAM capacity will improve a computer's speed and performance.

Resolution
The amount of fine detail portrayed in an image.

Scanner
A peripheral that scans a beam of light onto a flat surface and analyzes the reflected light in order to produce an image of the surface being scanned. Exact copies of drawings, images, and graphics can be produced in this way.

Scratchpad
A passive touch pad allowing text input, found on palm PDAs.

SD
'Secure digital'; one of the most popular solid-state memory cards available, used in a wide variety of digicams and PDAs.

(SSD)
'Solid-state hard drive'. An alternative to traditional HDDs, SSDs do not have any moving parts, and because of their lower power consumption, lack of noise, and lower heat generation they are increasingly being used in laptops.

Slate
A touch screen computer without a physical keyboard.

Tablet PC
A portable computer with a touch-sensitive screen (fingertip and/or stylus input).

Telescope
Traditionally, an optical device that uses a combination of lenses and/or mirrors to collect and focus light in order to view distant objects in greater clarity. Other forms of electro-magnetic radiation, such as radio waves, X-rays, infrared, and gamma rays can be detected and 'seen' using other types of telescopes.

TFT
See LCD.

Touch pad
A smooth area on a laptop enabling a fingertip to move the on-screen cursor.

Touch screen
A touch-sensitive screen that allows input to be made using a stylus or fingertip.

Transistor
A small semiconductor that can switch and modulate electric current.

UMPC
'Ultra-mobile PC'; a small (paperback-sized) slate computer.

Upgrade
Adding items of hardware or software to improve a computer's speed, performance, and/or appearance. Upgrades may involve physically opening a computer's case and attaching items to the motherboard.

USB
'Universal serial bus'; the most common type of connection between a computer and a peripheral device.

UT
'Universal time'; used by astronomers when recording observations.

WiFi

'Wireless fidelity'; a radio communications standard enabling devices to connect with each other and to the Internet.

Windows

An operating system (OS) developed by Microsoft, currently the most popular OS in terms of numbers. The latest Windows OS is called Vista.

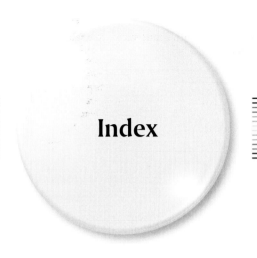

Index

Note: The locators in **Bold** refers to figures

A
Abacus, 6
Age of Enlightenment, 11
Aldrin, B., 22
America, pre-Columbian, 6
Antikythera mechanism, 5–6, **6**
Apollo Moon missions, 16, 21–22, 171
    Apollo Guidance Computer (AGC),
        21, **21**, **22**
    Apollo PGNS, **22**
Archaeoastronomers, xiv
Armstrong, N., 22, 98
Asterism, xiii, 112
Astrolabe, 7–8, **7**
Astrology, 8
Astronomical computing, mechanical
        devices
    Abacus, 6
    Antikythera mechanism, 5–6, **6**
    Astrolabe, 7–8, **7**
    Calendar wheel, 6
    Celestial globe, 7, **8**
    Cross staff, 7
    Orrery, 11
    Planisphere, 7–8
    Slide rule, 12
    Stonehenge, 4, **4**
Astronomical tables, 13, 14

Astronomy, computer programs,
        111–133
    Celestial tourism, 113
    Database and reference, 113
    Peripheral software & utilities, 114
    Planetarium, 34, 73, 112–113
Aurora, 142
Averted vision, 176

B
BAA Lunar Section, xi, **146**
BAA, xi, 165
Babbage, C., 12–13
Babylonia, ancient, 5
BBC, 45, 99
Bell Laboratories, 17–18, 42
Binoculars, 9, 175, 179, 180, 181
Birmingham Astronomical Society, 29
Birmingham Central Library, 14
Bletchley Park, 14–15
Blinn, J., **20**
Bluetooth, 79, 84, 92, 94, 205
Boeing, 18
Boyle, C (Earl of Orrery)., 11

C
Calendar wheel, 6
Cannon, A., 13

Cathode Ray Tube, *see* CRT
CCD images, *see* Digital imaging
Celestial globe, 7, **8**
China, ancient, xv
Cold War, 14–15
Color perception, 138–139
Colossus, 14–16
Columbia University, 14
Comet, xiv, 13, 25–26, 97–98, 113, 118
    Ikeya-Zhang, **197**
Computer-aided design (CAD), 18
Computers
    antivirus software, 38
    Apple computers, 30–32, 71
        Apple Macintosh, 29, **30**,
            32, 38–39
        Apple Newton, 86, **87**
        iMac, 32–33
        iPhone, 75, **76**
        MacBook, 33
        MacBook Pro, 33
        Macintosh Portable, 30
        MacWrite, 29
        Newton MessagePad, 85, 86, **87**
        Operating System, 33
        Powerbook, 30, 100
        PowerMac 7100, 32–33
    BASIC, 25–26, 112
    Central Processing Unit, *see* CPU
    Complex mechanical, 13
    CPU, 35, 73, 80, 81, 82, 83, 92–93, 206, 207
    Defragmenting, 37–38
    Diagnostics, 38
    Disk cleanup, 38
    Display, 25–26, 29, 45–54
        Active matrix, 52, 207
        Aspect ratio, 53
        Backlight, 50
        Brightness, 50, 52, 60
        Contrast, 52, 60
        Cost of, 50
        CRT, 17, 45, **46**, 47–48, 49–50, 71, 74,
            206, 207
        Dot pitch, 47–48, **48**
        Eye strain, 49
        Flicker, 49
        LCD, 49–50, **51**, 52, **52**, 54, 62, 65, 71,
            74, **75**, 76, 82, 85, 86–87, **89**,
            100, 105, 207
        Monochrome, 25–26, 29, 46, 71, 85, 86,
            98, 136, 138
        Passive matrix, 53, 73, 207
        Pixel, 27, 29, 34, 47–48, 50, **51**, 53, 86,
            89–90, 208
        Problem pixels, 53

        Resolution, 47, 48, 49
        Settings, 47
        Shadow mask, 45, 47, **48**, **49**
        Size, 49
        Submarining, 52
        Sub-pixels, 50, **51**, 52, 207
        SVGA, 46
        SXGA, 46
        Touchscreen, xiii, xviii, 74–77, **75**, 78,
            79, 85–87, 88–91, 92, 95, 169,
            170
        UXGA, 46
        VESA, 46
        VGA, 46, 83, 112
        Viewing angle, 52
        Widescreen, 53, **53**, **54**
        XGA, 46
Early electronic computer, 13, 14–17
Early generations of modern computer
    DAC-1, 18
    DEC Digital PDP-1, 18, **19**
    Ferranti Mark 1, 16
    IBM 7090, 17–18
    MADM, 16
    NORC, 16
    SSEC, 16
    SSEM('Baby'), 16
    'Super Foonly F-1', 19
    TX-2, 17
    Whirlwind, 17
Early PC graphics, 26–36
    Art Director, 27
    Degas, 27, **28**, **29**
Early PC models
    Atari 400, 45
    Atari 520, 27
    Atari 800, 45
    Atari Falcon, 27
    Commodore 64, 26, 45
    Commodore VIC-20, 45
    Olivetti PC-1, 46
    Sinclair Spectrum, 45
    Sinclair ZX81, 25, **26**, **27**, 45
    TS1000, 25
Expansion slot, 30
Flash memory, 60, 99
    See also, Computers, Memory card
Graphics
    Color, xviii, 27, 45, 46, 47, 48, 50, 52,
        62–63, 76, 85, 87, 92, 114,
        117, 135, 136, **137**, 139,
        140–141, 142, 149, 158, 172,
        177, 183, 185
    Color depth, 135, 136–138, 141
Early graphics programs

Boeing Man, 18
Catalog, 17
MacPaint, 29
Oscillon, 17
Sketchpad, 17
Spacewar!, 18, **19**
A Two Gyro Gravity Gradient
    Altitude Control System,
    17–18
Graphics card, 35, 46, 54
Graphics formats
    Bitmap graphics, 135–136, 141, 152,
        163, 165
    BMP, 59, 124, 138, 141, 142, 149,
        164, 172, 178, 180, 183, 206
    GIF, 141–142, 149, 206
    JPG, 141, 187, 198, 200, 206
    PNG, 141, 149, 206
    TIFF, 59, 141, 206
    TIFF-GIF conversion, **142**
    Vector graphics, 17, 85, 135, 141,
        147, 165–166
Graphics programs, 143–151
    Adobe Illustrator, 32
    Adobe Photoshop, 32, 34
    CorelDraw, 141, 144–145, **146**
    Corel PhotoPaint, 34, 35, **79**, 92, **140**,
        144–145, 152, 153–156, 172,
        177, 179, 182, 187, **189**, 198,
        199–200
    PaintShopPro, 8, 34
    Poser, 7, 33
    Terragen, 33
Grayscale, 59, 76, 139, **140**, 141, 142,
    157, 158, 172, 177, 179, 181,
    183, 187, 194–196, 198, 200
Properties, commands and tools
    Blend, 153, 174, 178, 182, 187, 198
    Blur, 69, 144, 159, 162, 172, 177, 178,
        180, 182
    Brighten, 158, 174, 195
    Brightness, 149, 153, 157, 195–196
    Clone, 153–155, **156**
    Contrast, 68–69, 144, 149, 153, 157,
        195–196, 199–200
    Crop, 152, 158, 165, 178, 187,
        194–196, 199–200
    Darken, 158, 174
    Despeckle, 160
    Erase, 148
    Fill, 29, 44–45, 153–154, **156**, 163,
        172, 174, 182, 187, 198
    Flip, 164, 165, 179
    Gaussian blur, 159, **159**, **160**, **195**
    Icon, 152, 153

Layers, 144, 148, 163–164, **163**, 179,
    180, **181**, 182, 198
Mask, 163, 179
Median filter, 159–160, **160**, **161**,
    173–174, 179, 181, 183,
    187, **195**
Moiré, 161, **161**, 162
Noise, 149, 159, 161
Paintbrush, 153, 174, 180
Paint roller, xviii
Paint, xviii, 29, 143, 153, **153**, **154**,
    178, 180, 200
Pen, 29, 153
Resample, 164, 172, 181, 183, 198
Rotate, 164, 165
Screen capture, **150**, 151–152, 170
Shade, 29, 177, 182, 187
Shape, xviii, 29, 85, 149, 153, 154,
    165–166, 180, 205
Sharpen, 68–69, 153, 159, 162,
    174, 179
Smear, 153, 174, 187, 198
Smudge, 144, 153, 177, 178,
    182, 198
Speckle, 142, 160–162
Spray, xviii, 29, 153, 154, 159
Text, 85, 149, 165, 170, 175
Texture, xviii, 69, 144, 147, 153
Tone curve, 158, **158**, **160**, **161**, 179,
    187, 199–200
Transparency, xviii, 153,
    154–155, 200
Undither, 153, 198
Unsharp mask, 69, **69**, 149, 162,
    162, 181
Zoom, xviii, 112, 113, 129, 135, **136**,
    169, 170, 172, 183
GUI, 29, 30–32, 38–39, 190, 207
Handheld PC (H/PC), xviii, 88–89, 91, 100,
    101, 142, 149, 207
    Battery, **102**
    CF slot in, **102**
    Jornada 77, **89–90**, 101, **102**, **142**, 720
Hard disk drive, 16, 24, 29, 30, 34, 37, 38,
    38, 45, 67–68, 74, 79, 80, 81,
    82, 83, 97, 100, 107, 173, 181,
    187, 207, 208, 209
    CF Microdrive, 103–105
    Data backup, 37, 97, 175
    PC Card HDD, 100
    Shock-mounted HDD, 81–82
    SMART, 38
    SSD, 209
    USB HDD, 103–105
Hard drive, *see* Hard disk drive

Computers (*cont.*)
  HDD, *see* Hard disk drive
  Hot keys, 117, 152
  H/PC, *see* Handheld PC
  Human computer, 13
  IDE cable, 38
  Integrated circuit, 22–23
  Laptop, 3, **8**, 30, 33, 34, 53, 71–74, 78–79,
        89–90, 95, 100, 106, 117, 207,
        209
    Keyboard, 72
    Laptop models
      Compaq Presario 1200, 73
      Macintosh portable, 71
      NEC UltraLite, 71
      Osborne, 1, 71
      Radio Shack TRS-80 Model, 71, 100
    Operating temperature, 73
    Thinnest, 33
    Touchpad, 72, **72**, 209
    Tough laptop, 74
    Trackball, 72
    Types, 71–72
  Magnetic core memory, 17, 21, 22–23
  Miniaturization, 24
  Monitor, *see* Display
  Notebook, *see* Laptop
    Microsoft OS
      MS-DOS, 30–32
      Windows 2000, 33
      Windows 3.0, 30–32
      Windows 3.1, 33
      Windows 95, 33
      Windows 98, 33
      Windows ME, 33
      Windows Vista, 33, 34, 35
      Windows XP, 33, 34
    Operating System (OS), 30–32, 33, 37,
        38, 208
  Palmtop, *see* Handheld PC
  PDA
    Advantages of, 93–95
    Audio, 94
    Battery, **8,** 92–93, **102,** 170, 175
    Brightness, 170, 176, 185
    Camera, 63, **64**
    Features of, 90–91, **90**
    Keyboard, 89, **93**
    Palm PDA, 86–88
      Graffiti scratchpad, 86–87
      Palm Operating Systems
        Nova, 87–88
        Palm OS, 86–88
      Palm PDA models
        Palm III, 87

  PalmPilot, 86
  Palm Tungsten E, **89**
  Palm Tungsten T3, 87
  Sony Clié NZ90, 87
Pocket PC (P/PC)
  P/PC models
    iPAQ 6300, 77, **90, 91**
    iPAQ, 92
    Jornada 540, 77, 92, 95, 100–101,
        102, 133, 138, 182, **191**
    Samsung i900, 75, **76**
    SPV M2000, 92, **93, 94,** 142, 172,
        177, 179, 180
    XDA, **75, 105**
    XDA Orbit, 78, 92, **94**
  P/PC Operating Systems
    Handheld PC 2000, 91, 142, 149
    Pocket PC 2000, 91, 149
    Pocket PC 2002, 91, 149
    WinCE, 86–87, 88–89, 91
    Windows Mobile 2003, 91, 149
    Windows Mobile 2003 SE, 91
    Windows Mobile 5.0, 91
    Windows Mobile 6.0, 91
  P/PC Processors, 91
Screen size & format comparison,
    90, **91**
Stylus, 74, 75, **75,** 76–78, **77,** 85–86, 94,
    95, 111
Touchscreen
  Reflections, 76
  Resistive touchscreen, 74
  Sensitivity, 75–76
  Surface capacitive touchscreen, 75
Peripherals
  Digital notepad, 44–45, **44**
  Floppy disk/drive, 29, 71, 97–98, 99,
      106, 206
    SmartMedia adaptor, **103**
    USB floppy drive, 98
  Graphics tablet, 41–43, **43,** 206
    Active tablet, 43
    Eraser, 42–43
    Koala Pad, 42
    Passive tablet, 42
    Powered stylus, 43
    Pressure levels, sensitivity, 42
    Wacom Graphire, 42
  Keyboard, 26, 27, 29, 30, 35, 38–39, 40,
      42, 84, 152, 207, 209
  Mouse, 26, 38–41, **39, 40**
    Cordless, 40
    Laser, 41
    Mechanical encoder, 39
    Optical, 39

Opto-mechanical, 39
Scrolling wheel, 40
Printer, 30, 55–57, **56**, **57**
   DMP, 55
   Inkjet, 55, **56**
   Inkjet refill, 55
   Laser, 55–56, **57**
Scanner, 57–60, **59**
   Flatbed, 57–59
   Handheld, 58
   Scanning formats, 59
   Scanning tips, 59–60
   Software, 60
Trackball, 30, 39
Personal Digital Assistant, *see* PDA
Photograph, first processed, 17
PIM, 85
Portable data storage, 97–107
   CD-ROM, 98, 99, 100, 121
   DVD-ROM, 113
   Memory card, 44–45, 66, 89, 99–107,
      133, 172, 175, 205, 209
      CF, 100–101, **102**, 103–105, **104**, 205
      Memory Stick, 103, **104**, 106–107
      Memory Stick Duo, **104**
      Micro SD, **105**
      PC Card, 100
      PCMCIA, *see* PC Card
      SD, 86, **101**, 103, **104**, **105**, 209
      SDHC, 103
      SmartMedia, 101, **103**
      xD, 103, 106–107
   Memory card reader, 84, **103**, **106**
P/PC, *see* Pocket PC
PPC, *see* Pocket PC
Processing speed, 24
Processor, *see* CPU
PS/2 port, 40
Punched card, 14
RAM, 16, 21, 24, 29, 30, 34, 35, 36, **36**
Random Access Memory, *see* RAM
REV drive, 103–105
Serial port, 42
Slate, 79
Specifications, 34, 35
Stored program, 16
Stylus, xviii, 27, 41–43, **43**
Tablet PC, xviii, 34
   Active stylus, **43**, 81
   Audio, 81, 83
   Battery, 81, 82, 83
   Brightness, 185
   Comparisons, 80–84
   Keyboard, 78, 79, 84
   Models

   Fujitsu Stylistic 3400, 80, **80**
   Fujitsu Stylistic ST5111, 83, **83**
   Hammerhead HH3, 81, **81**
   Samsung Q1, 84, **84**
   Viewsonic V1100, 82, **82**
  Resistive panel, 80, 83
  Stylus, 78
  Tough tablet PC, **81**
  UMPC, xviii, 3, 34, 78–79, **84**
Upgrade, 36
USB port, 32
Video card, 46–47
Virus, 33, 38
Zip drive, 98
CPU, 35, 36, 73, 80, 81, 82, 83, 84, 92, 207
Crookes, W., 45
Cross staff, 7
CRT, 17, 45, 46, 47, 48, 49, 50
Cybersketching
   Basic drawing tools, 152–166
   'Cybertemplate', *see* Digital image template
   Digital image template, 61, 165,
      193–201, **197**
   Freehand, **189**
   Halftoning, 201–203
   Observational field drawings, 167–203
   Programs, 143–152
   Revitalizing conventional drawings,
      201–203
   Saving to PDA, 174–176
   Simultaneous cybersketching, 190
   Size of, 169, **173**
   On Tablet PC / UMPC, 187, **188**
   Techniques, xviii
   Tutorials
      Colored double stars on PDA, 183,
        184, **185**
      DSO, digital image template,
        193–194, **197**
      Freehand DSO on PDA, 176–181,
        178, **180**
      Freehand Jupiter on PDA, 181–183
      Freehand lunar sketch on PDA,
        171–174, **173**, **174**
      Freehand lunar sketch on Tablet PC,
        187, **188**, **189**
      Lunar sketch, digital image template,
        193–194, **195**, **197**
   Using PDA, 170–183
   Using Tablet PC / UMPC, 183–189

D
Dark adaptation, 175–176
D'Arrest, H., 11
Da Vinci, L., 9

Deep Sky Object (DSO)
   Crab Nebula (M1), **130**
   Dumbbell Nebula (M27), **159**
   M15 (globular cluster), **118, 162, 186**
   M3 (globular cluster), **191**
   M44 (open cluster), **148**
   M49 (galaxy), **149**
   M65, M66 & NGC 3628 (galaxies), **150**
   M6 (open cluster), **181**
   M81 & M82 (galaxies), **121**
   NGC 6709 (open cluster), **161**
   NGC 6826 (planetary nebula), **127**
   NGC 7331 (galaxy), **178, 180**
   Orion Nebula (M42), **129, 200**
   Ring Nebula (M57), **192**
Desktop Publishing, *see* DTP
Digicam, 53, 60, 61, 62, 63, 66, 100, 195,
      197, 205
Digital image template, 61, 165, 193, 194, 197,
      198, 200
Digital imaging, 60–69
   Afocal, 61, **61**, 63, 65–66, 67, 94, 194
   AVI, 68–69, 124–125, **191**
   Camcorder, 65–67, **66**
   CCD, xvii, 24, 58, 60, 67, 68, 97, 114,
      122, **162**, 189, 190, 192,
      193, 194
   Digicam, 53, 61–63, **61, 62**, 66, 94, 97, 100,
      125, 195–196
   Digital camera, *see* Digicam
   Digital SLR, *see* DSLR
   DSLR, 61, 65, **65**, 140, 198, 199
   Eyepiece projection, 67
   LCD screen, 62, 74
   Mobile phone camera, 63, **64**
   PDA camera, 63, **64**
   Prime focus, 65, 67, **68**, 198
   Stacking, 67, 122, 124
   T-mount adapter, 65
   Video camera, 190–193
   Vignetting, 63
   Webcam, 34, 61, 66, 67–69, 73, 95, 190
   Zoom, 63
Dimond, T L, 42
Disney Studios, 18–19
Display, 6, 7, 25, 39, 40, 47, 48, 49, 50, 54, 61, 73,
      74, 80, 81, 90, 105, 112, 121,
      139
Drawing (conventional)
   Digital revitalization, 199, **200, 201–203**
   Observational, 55
   Reasons for, xvii
   Using digital image template, 193–194
   Using video camera, 190–193
   From video loop, 67

DTP, 30, 32
Dunhuang manuscript, **xvi**
DVD-video, 54

E
Eckert, W., 14, 16
École Polytechnique Fédérale de Lausanne, 39
Englebart, D., 39
English, William, 39
ENIAC, 15–16, **15**
Epicycle, 5
Eratosthenes, 5
Eudoxus, 5
Eugene, Prince of Savoy, 11
Eyepiece, xvii, xviii, 60, 61, 63, 65–66, **66**, 67,
      73, 94, 114, 138–139, 169,
      171, **173**, 177, 181, 190

F
Fairchild Semiconductor, 22–23
Filters, 50, 58, 139, 148, 149, 156–158, 164
Flowers, T., 14–15
France, xiii
Fraunhofer (crater), 172, **173, 174**

G
Galilei, Galileo, 10–11
Galle, J., 11
Gauss (crater), 198
GCHQ, 14–15
General Motors, 18
Geocentric Universe, 5
Gilbert, W., 9, **10**
Gods, sky, 3
Goldberg, H., 41–42
Graham, G., 11
Graphical User Interface (GUI), 30–32
Greece, ancient, 5
Grego, J., **145**
GUI, *see* Graphical User Interface (GUI)

H
Handheld PC, 89, 91, 142, 149
Hansfel, G., 42
Hard disk drive, 36, 37, 97
Hardy, D., 27, 32–33
Harriot, T., xv, 167
Harvard College Observatory, 13
Heliocentric Universe, 5
Herschel, C., 13
Herschel, W., 11
Hipparchus, 5, 7
Hollerith, H., 14
Holt, D., 190
Holzman, B., 19

I

IBM, 14, 16, 17–18
Intel, 23

J

Jansen, S., 9
JPL Computer Graphics Laboratory (CGL), 19
Jupiter, xiii, **xv**, 11, 19, 94, 113, 119, **119**,
         121, 124, 125, 138–139,
         142, 170
   Drawing on PDA, 181–183
   Satellites of, 11, 17–18, 121, 169

K

Kepler, J., 8
Kilby, J., 22
Krieger, J., 167–169, 194

L

Laposky, B., 17
Laptop, 8, 33, 71, 72, 73, 74, 89, 100, 117
Lascaux caves, xiii, **xiv**
LCD, 49, 50, 51, 52, 53, 61, 62, 65, 71, 75, 76, 85,
      87, 100
Leavitt, H., 13
LED, 41
Leibniz, G., 12
Lens, 8
Le Verrier, U., 11
Light pen, 17, **18**
Linux, 33
Lippershey, H., 8
Lisberger, S., 18
Logitech, 39
Longshaw, N., 201
Lyra, **27**

M

Manchester Museum of Science &
        Industry, 16
Mannheim, A., 12
Mars, 29, **31**, **33**, **84**, 113, 115, 117, 120,
        120, 125, 139, 142, **142**, 159,
        160, 166, 168, **168**, 170, **195**,
        196, **202**
Masuoka, F., 99
Maury, A., 13
Mercury, xi, 5, 166, 170, **186**
Meteor, 126
Metius, A., **8**
Metius, J., 9
Microsoft programs
   ActiveSync, 35
   MS-DOS, 30
   Mystify, 17

WinCE, 86, 88, 91
Windows 2000, 33
Windows 3.0, 30
Windows 3.1, 33
Windows 95, 33
Windows 98, 33
Windows Aero, 35
Windows Experience Index, 35
Windows ME, 33
Windows Paint, 80
Windows Vista, 33, 34, 35
Windows XP, 33, 34
Milky Way, 11
MIT, 17, 18
Mitchell, M., 13–14
MIT Lincoln Laboratory, 17
Mondrian, P., 26
Moon, xiii, xiv, xv, **xv**, xvi, xvii, 3, 5, 8, 9, 10, **10**,
        11, 12, 16, 21, 22, **23**, **28**, 61, 62,
        63, 64, 66, 67, 68, 94, 113, **115**,
        119, 120, **128**, 129, **130**, 132,
        138, 139, 142, **145**, **146**, **168**,
        168, 169, 171, 172, 187, 193,
        193, 194, 195, 196
   Astronauts, 12
   Craters, 11, 25
   Distance and size, 5
   Earthshine, 9, 10, **10**
   Eclipse, **116**, **143**
   Ephemeris, 16
   Features, xv, xvii, 9–11
      Alphonsus (crater), **202**
      Archimedes (crater), **156**
      Cassini (crater), **168**
      Clavius (crater), **157**
      Euler & Mons Vinogradov, **193**
      Gauss (crater), **195**
      Jansen (crater), 9
      Lansberg (crater), **32**
      Lippershey (crater), 8
      Mare Crisium, 9
      Mare Nubium, 9
      Mare Tranquillitatis, 9, 22, 98
      Metius (crater), **8**
      Oceanus Procellarum, 9
      Petavius (crater), **196**
      Plato (crater), **160**
      Plinius(crater), **188**, **189**
      Statio Tranquillitatis, 22
      Steinheil (crater), **159**
      Stevinus (crater), **191**
   Map, 9–10, **10**, **23**, **28**
   Observing, xvi, xvii
   Phases of, xiii, xv
   Surface of, 11

Moore, G., 23
Moore's Law, 23–24, 34
Morgan, P., **202**

N
Nagler, 193
NASA, 19
Nebula, xvii, 67, 113, **129**, **130**, 139, **159**, 175,
176, 177, 192, **192**, 198, 200,
200
Neptune, 11, 19
Newman, M., 14
NGC, 113, 123, 126
Noyce, R., 22, 23

O
Observational field drawings, 167
'Optick tube', xv, 8
Orrery, 11–12, **12**
Ottoman Empire, 8
Oughtred, W., 12

P
Paris Observatory, 168
Pascal, B., 12
PC, *see* Computers
PC Card, 100
PDA, xv, 10, 34, 35, 63, 76, 77, 78, 85, 89, 92, 94,
95, 105, 114, 133, 141, 149,
161, 163, 164, 170, 171, 173,
175, 176, 177, 181, 182, 183,
191
PDF, 90, 165
Pencil sketching, 163
Penny, J., 29
Personal Computer, *see* Computers
Personal Digital Assistant, *see* Computers,
PDA
Photographic techniques, 27
Photography, xv
Pickering, Edward, 13
PIM, 85, **86**, 208
Planetary motion, laws of, 8
Planisphere, 7–8
Pocket PC, 86, 90, 91, 92, 124
Ptolemaeus, C., 5

R
Radar, 17, **18**
RAM, 16, 21, 23, 24, 25, 29, 32, 34, 35, 36, 84,
85, 92
RAND, 42
Retrograde motion, 5
Rowley, J., 11
Russell, S., 18

S
SAGE, 17
Sale, T., 15
Saturn, xviii, 11, 19, **20**, **88**, 94, **116**, 119, **119**,
125, **127**, 136, **136**, **151**, 152,
166, **166**, 170
Science Museum, London, 13
Silicon chip, 23–24
Sketching, *see* Drawing (conventional)
Slide rule, 12–13
Smudge, xvii, 57, 144, 153, 177, 178, 179,
182, 198
Solar System, 11, 25
Stanford Research Institute, 38
Star map, **xvi**, 7, 25, **27**
Stellar spectra, 13
Stippling, 201, 203
Stonehenge, 4, **4**
Sumeria, ancient, 6
Sun, xiv, 3, 4, 5, 7, 11, 26, 62, 63, 97, 138, 171,
172, 208
Sutherland, I., 17

T
Telescope, 209
Binocular eyepiece (binoviewer), 175,
177, 187
Eyepiece, xv, xvi, xvii, xviii, 60, 61, 62, 63,
65, 66, 67, 73, 81, 94, 114, 138,
139, 142, 152, 164, 165, 167,
169, 171, 173, 176, 177, 181,
189, 190, 192, 193, 194, 196,
206
Meade LX90, 177, 182, 187, 191
Newtonian, xi, 191, 192, 197
Refractor, xi, 61, 167, 168, 192
Television, 45
Texas Instruments, 22
TFT, *see* LCD
Tomorrow's World, 99
Transistor, 17, 23, 24, 100, 207, 209
Tron, 18–19
Turing, A., 14

U
University of Aston in Birmingham, 29
Uranus, 11, 19, 125
US Air Force, 17, **18**
US Army Ballistic Research Laboratory, 15
US Census Bureau, 14
US Naval Observatory, 13, 14

V
Valves, electronic, 14
Venus, xi, 5, 13, 166, 170, **201**
Vermeer, J., 8

Video adapter card, 49
Visual observing, reasons for, xvii
Voyager space probes, 19, **20**

W
Wacom, 43

Whitney, J., 17
WiFi, 79, 83, 84, 92, 94, 95, 100, 114, 187, 210
World War II, 14–15, 17

Y
Yardley's School, 97

Printed in the United States